Oil and Gas Well Cementing for Engineers

Oil and Gas Well Cementing for Engineers

Baghir A. Suleimanov
Oil and Gas Scientific Research Project Institute
State Oil Company of Azerbaijan Republic (SOCAR)
Baku, Azerbaijan

Elchin F. Veliyev
Oil and Gas Scientific Research Project Institute
State Oil Company of Azerbaijan Republic (SOCAR)
Baku, Azerbaijan

Azizagha A. Aliyev
Oil and Gas Scientific Research Project Institute
State Oil Company of Azerbaijan Republic (SOCAR)
Baku, Azerbaijan

This edition first published 2023
© 2023 John Wiley & Sons Ltd

All rights reserved. No part of this publication may be reproduced, stored in a retrieval system, or transmitted, in any form or by any means, electronic, mechanical, photocopying, recording or otherwise, except as permitted by law. Advice on how to obtain permission to reuse material from this title is available at http://www.wiley.com/go/permissions.

The right of Baghir A. Suleimanov, Elchin F. Veliyev, and Azizagha A. Aliyev to be identified as the authors of this work has been asserted in accordance with law.

Registered Offices
John Wiley & Sons, Inc., 111 River Street, Hoboken, NJ 07030, USA
John Wiley & Sons Ltd, The Atrium, Southern Gate, Chichester, West Sussex, PO19 8SQ, UK

For details of our global editorial offices, customer services, and more information about Wiley products visit us at www.wiley.com.

Wiley also publishes its books in a variety of electronic formats and by print-on-demand. Some content that appears in standard print versions of this book may not be available in other formats.

Trademarks: Wiley and the Wiley logo are trademarks or registered trademarks of John Wiley & Sons, Inc. and/or its affiliates in the United States and other countries and may not be used without written permission. All other trademarks are the property of their respective owners. John Wiley & Sons, Inc. is not associated with any product or vendor mentioned in this book.

Limit of Liability/Disclaimer of Warranty
In view of ongoing research, equipment modifications, changes in governmental regulations, and the constant flow of information relating to the use of experimental reagents, equipment, and devices, the reader is urged to review and evaluate the information provided in the package insert or instructions for each chemical, piece of equipment, reagent, or device for, among other things, any changes in the instructions or indication of usage and for added warnings and precautions. While the publisher and authors have used their best efforts in preparing this work, they make no representations or warranties with respect to the accuracy or completeness of the contents of this work and specifically disclaim all warranties, including without limitation any implied warranties of merchantability or fitness for a particular purpose. No warranty may be created or extended by sales representatives, written sales materials or promotional statements for this work. The fact that an organization, website, or product is referred to in this work as a citation and/or potential source of further information does not mean that the publisher and authors endorse the information or services the organization, website, or product may provide or recommendations it may make. This work is sold with the understanding that the publisher is not engaged in rendering professional services. The advice and strategies contained herein may not be suitable for your situation. You should consult with a specialist where appropriate. Further, readers should be aware that websites listed in this work may have changed or disappeared between when this work was written and when it is read. Neither the publisher nor authors shall be liable for any loss of profit or any other commercial damages, including but not limited to special, incidental, consequential, or other damages.

Library of Congress Cataloging-in-Publication Data

Names: Suleimanov, Baghir, author. | Veliyev, Elchin F., author. | Aliyev, Azizagha, author.
Title: Oil and gas well cementing for engineers / Baghir A. Suleimanov, Elchin F. Veliyev, Azizagha A. Aliyev.
Description: Hoboken, NJ : Wiley, 2023. | Includes index.
Identifiers: LCCN 2023017127 (print) | LCCN 2023017128 (ebook) | ISBN 9781394164851 (hardback) | ISBN 9781394164868 (adobe pdf) | ISBN 9781394164875 (epub)
Subjects: LCSH: Oil well cementing.
Classification: LCC TN871.2 .S83 2023 (print) | LCC TN871.2 (ebook) | DDC 624.1/833–dc23/eng/20230427
LC record available at https://lccn.loc.gov/2023017127
LC ebook record available at https://lccn.loc.gov/2023017128

Cover Design: Wiley
Cover Image: © Edelweiss/Adobe Stock Photos

Set in 9.5/12.5pt STIXTwoText by Straive, Pondicherry, India

Contents

Foreword *xiii*
Introduction *xv*

1 Theoretical and Practical Aspects of Well Cementing *1*
1.1 Oil Well, Its Elements, and Construction *1*
1.2 Objectives of Well Cementing *5*
1.3 Primary Cementing *9*
1.3.1 Single-Stage Cementing with Two Plugs *10*
1.3.2 Two-Stage (Two-Cycle) Cementing *11*
1.3.3 Basket Cementing *12*
1.3.4 Liner Cementing *13*
1.3.5 Reverse Cementing *14*
1.3.6 Cementing Plugs *14*
1.4 History of Oil Well Cementing Technology Development *16*

2 Composition and Classification of Portland Cement *19*
2.1 Chemical Composition *19*
2.2 Portland Cement Manufacturing *22*
2.3 API (American Petroleum Institute) Classification of Portland Cement *24*
2.4 GOST (Russian: ГОСТ) Classification of Portland Cement *29*

3 Cement Additives *31*
3.1 Introduction *31*
3.2 Accelerators *32*
3.3 Retarders *36*
3.3.1 Lignosulfonates *37*

3.3.2	Hydroxycarboxylic Acid	38
3.3.3	Saccharide Compounds	38
3.3.4	Cellulose Derivatives	38
3.3.5	Organophosphonates	39
3.3.6	Inorganic Compounds	39
3.4	Extenders	39
3.4.1	Clays	40
3.4.2	Sodium Silicate	43
3.4.3	Pozzolans	43
3.4.3.1	Diatomaceous Earth (Kieselgur)	44
3.4.3.2	Fly Ash	44
3.4.3.3	Lightweight Cementing Slurries	45
3.4.3.4	Silica (Silicon Dioxide, Quartz)	45
3.4.4	Lightweight Particles	46
3.4.4.1	Expanded Perlite	46
3.4.4.2	Gilsonite (Asphaltum)	46
3.4.4.3	Powdered Carbon	47
3.4.4.4	Microspheres	47
3.4.5	Gas Based Extenders	48
3.4.5.1	Nitrogen	48
3.5	Weighting Agents	48
3.5.1	Ilmenite (Iron Titanium Oxide)	49
3.5.2	Hematite	49
3.5.3	Hausmannite	49
3.5.4	Barite	50
3.6	Dispersants	50
3.7	Fluid Loss Agents	53
3.7.1	Particulate Materials	54
3.7.2	Water Soluble Polymers	54
3.8	Lost Circulation Prevention Agents	55
3.9	Special Cement Additives	55
3.9.1	Antifoaming Agents (Defoamers)	55
3.9.2	Strengthening Agents	56
3.9.3	Radioactive Tracers	56
3.9.4	Mud Decontamination	57
4	**Special Cement Systems**	**59**
4.1	Thixotropic Cement	59
4.2	Expansive Cement	61
4.3	Freeze-Protected Cement	62
4.4	Salt-Cement Systems	63

4.5	Latex-Cement Systems	*64*
4.6	Corrosion-Resistant Cement	*65*
4.7	BFS Systems	*66*
4.8	Engineered Particle-Size Distribution Cements	*67*
4.9	Low-Density Cements	*69*
4.9.1	Foamed Cement	*69*
4.10	Flexible Cement	*70*
4.11	Microfine Cements	*71*
4.12	Acid-Soluble Cements	*72*
4.13	Chemically Bonded Phosphate Ceramics	*72*
4.14	Special Cement Systems	*73*
4.14.1	Nonaqueous Cement Systems	*73*
4.14.2	Storable Cement Slurries	*73*

5 Cementing Equipment *75*

5.1	Surface Equipment	*75*
5.2	Casing Types	*84*
5.2.1	Conductor Casing	*86*
5.2.2	Surface Casing	*86*
5.2.3	Intermediate Casing	*86*
5.2.4	Production Casing	*86*
5.2.5	Liner	*87*
5.3	Technical Characteristics of Casing	*88*
5.3.1	Steel Grades	*88*
5.3.2	Strength Characteristics of Casing	*91*
5.3.3	Weight Per Unit Length of Tube	*94*
5.3.4	Connection Types of Casing	*95*
5.4	Casing Hardware	*96*
5.4.1	Casing Shoe	*96*
5.4.2	Check Valve	*99*
5.4.3	Centralizer	*100*
5.4.4	Turbulator and Scratcher	*102*
5.4.5	Cementing Plugs	*103*
5.4.6	Cementing Head	*104*
5.4.7	Screening Devices and Cement Baskets	*105*
5.5	Remedial Cementing Equipment	*106*
5.5.1	Cased – Hole Remedial Cementing Equipment	*106*
5.5.1.1	Packers for Squeeze Cementing Operations in Cased Wells	*106*
5.5.1.2	Wellbore Tools for Tubing Pressure Testing and Pressure Equalization in the String and Annulus	*108*
5.5.2	Open Hole Remedial Cementing Equipment	*108*

6	**Primary Cementing** *109*	
6.1	Planning *109*	
6.1.1	Depth and Design of the Well *109*	
6.1.2	Reservoir Conditions *113*	
6.1.2.1	Pressure *113*	
6.1.2.2	Temperature *113*	
6.1.3	Drilling Mud Parameters *114*	
6.2	Slurry Selection *114*	
6.2.1	Density *114*	
6.2.2	Compressive Strength and Mechanical Properties *115*	
6.2.3	Formation Temperature *115*	
6.2.4	Cement Slurry Additives *116*	
6.2.5	Cement Slurry Design *116*	
6.3	Theoretical Basis of Mud Displacement *117*	
6.3.1	Preparing the Well for Running Casing *118*	
6.3.2	Theoretical Basis for Assessing Circulation and Displacement Efficiency *118*	
6.3.3	Conditioning the Drilling Mud *120*	
6.3.4	Drilling Mud Displacement *122*	
6.4	Methods of Well Cementing *124*	
6.4.1	Cementing Through Drill Pipes *125*	
6.4.2	Cementing Through Small Diameter (Macaroni) Tubing *126*	
6.4.3	Single-Stage Cementing *127*	
6.5	Multistage Cementing *128*	
6.5.1	Standard Two-Stage Cementing *128*	
6.5.2	Continuous Two-Stage Cementing *131*	
6.5.3	Three-Stage Cementing *132*	
6.6	Liner Cementing *133*	
6.7	Critical Factors in Cementing Operations *138*	
6.7.1	Volume of Cement Slurry *138*	
6.7.2	Displacement of Cement Slurry *138*	
6.7.3	Well Temperature *139*	
6.7.4	Well Pressure *139*	
7	**Remedial Cementing** *143*	
7.1	Plug Cementing *144*	
7.1.1	Plug Cementing Techniques *144*	
7.1.1.1	The Balance Method *145*	
7.1.1.2	Cement Plug Installation Using a Dump Bailer *145*	
7.1.1.3	Cement Plug Installation Using the Two Plugs Method *146*	
7.1.1.4	Cement Plug Installation with the Use of Coiled Tubing *146*

7.1.2	Plug Cementing Equipment	*147*
7.1.2.1	Bridge Plug	*147*
7.1.2.2	Tailpipe or Stinger	*148*
7.1.2.3	Diverter	*148*
7.1.2.4	Mechanical Separators	*148*
7.1.3	Slurry Design	*148*
7.1.4	Plug Cementing Evaluation	*149*
7.2	Squeeze Cementing	*149*
7.2.1	Squeeze Cementing Technologies	*152*
7.2.1.1	Classification of Squeeze Cementing Technologies According to Squeezing Pressure	*152*
7.2.1.2	Classification of Squeeze Cementing Technologies Depending on the Method of Injection of Cement Slurry	*153*
7.2.1.3	Classification of Squeeze Cementing Technologies According to the Method of Operation	*154*
7.2.2	Slurry Design	*155*
7.2.2.1	Fluid Loss	*156*
7.2.2.2	Rheology	*157*
7.2.2.3	Thickening Time	*157*
7.2.3	Design and Execution of Squeeze Cementing Operations	*157*
7.2.3.1	Determination of the Cement Slurry Volume	*157*
7.2.3.2	Spacer, Washer, and Displacing Fluids	*158*
7.2.3.3	Determination of Well Injectivity	*159*
7.2.3.4	Main Procedures for Squeeze Cementing Operations	*159*
7.2.4	Analysis and Evaluation of the Squeeze Cementing Job	*160*
8	**Cement Job Evaluation**	*163*
8.1	Hydraulic Testing	*164*
8.1.1	Pressure Test	*164*
8.1.2	Inflow Test	*167*
8.2	Temperature Log	*167*
8.3	Radioactive Logging	*169*
8.3.1	Pulsed Neutron Logging	*170*
8.3.1.1	Oxygen-Activated Neutron Gamma Method	*171*
8.4	Acoustic Logging	*171*
8.5	Types of Logging Tools	*176*
8.5.1	Cement Bond Log (CBL)	*176*
8.5.2	Radial Acoustic Cement Meter	*177*
8.5.3	Multiple Pad Sonic Tool	*177*
8.5.4	Ultrasonic Tool	*177*

9	**Laboratory Testing and Evaluation of Well Cements**	*179*
9.1	Preparation of Cement Slurry	*180*
9.2	Test Methods of Cement Slurries	*181*
9.2.1	Density	*181*
9.2.2	Thickening Time	*182*
9.2.3	Fluid Loss	*186*
9.2.4	Free Water	*187*
9.2.5	Sedimentation Test	*188*
9.2.6	Rheological Measurements	*188*
9.2.6.1	Flow Types	*188*
9.2.6.2	Laminar Flow	*189*
9.2.6.3	Turbulent Flow	*190*
9.2.6.4	Basic Rheological Concepts	*190*
9.2.6.5	Rheological Models	*191*
9.2.6.6	Newtonian Fluids	*192*
9.2.6.7	Non-Newtonian Fluids	*192*
9.2.6.8	Power-Law Model	*193*
9.2.6.9	The Bingham Model	*193*
9.2.6.10	Herschel–Bulkley Model	*194*
9.2.7	Static Gel Strength (SGS)	*196*
9.2.8	Flowability of Cement Slurries	*197*
9.3	Test Methods of Cement Stone	*199*
9.3.1	Mechanical Strength of Cement	*199*
9.3.2	Destructive Test (Compressive Strength)	*199*
9.3.2.1	Non-destructive Test (Ultrasonic Measurement)	*200*
9.3.3	Expansion and Shrinkage	*200*
9.3.4	Gas Migration	*202*
9.3.5	Cement Stone Permeability	*202*
9.3.6	Thermophysical Properties of Cement	*202*
9.3.6.1	Thermal Conductivity	*203*
9.3.6.2	Coefficient of Linear Thermal Expansion	*203*
9.4	Laboratory Evaluation of Spacers and Washers	*204*
9.4.1	Compatibility of the Buffer/Washer Fluid with the Drilling Fluid and Cement Slurry	*204*
9.4.2	Efficiency of Wellbore Cleaning with Washer Fluid	*204*
9.5	Chemical Analysis of Mix Water	*205*
10	**Typical Calculations for Well Cementing**	*207*
10.1	Slurry Preparation Calculations	*207*
10.1.1	Specific Gravity of Cement Slurry	*208*
10.1.2	The Concept of Absolute and Bulk Volumes	*208*

10.1.3	Additive Concentration Calculation	*209*
10.1.4	Density and Yield of the Slurry	*210*
10.1.5	Special Additives	*212*
10.1.5.1	Sodium Salt	*212*
10.1.5.2	Fly Ash	*214*
10.1.5.3	Bentonite	*216*
10.1.5.4	Weighting Agents	*218*
10.2	Primary Cementing Calculation	*218*
10.2.1	Volume of Cement Slurry	*221*
10.2.2	Volume of Displacing Fluid	*221*
10.2.3	Pressure to Place the Cement Plug on the Stop Collar	*222*
10.2.4	Buoyancy	*223*
10.3	Remedial Cementing Calculations	*225*
10.3.1	Plug Cementing Calculations	*225*
10.3.2	Squeeze Cementing	*229*

Annex. Conversion Tables *237*
Recommended Literature *245*

Index *247*

Foreword

Well cementing is largely integrated into the well construction process, being one of the two most common operations involved. It would be wrong to consider this process without taking into account the processes preceding and following this operation because the use of the cementing materials is not limited to drilling wells, but it is also widely used in the operation and production processes, being an inseparable part of them. Nowadays the number of fields at later stages of development is increasing year by year, and it is not difficult to assume that the number of operations related to well workover and bottomhole zone stimulation is rising almost proportionally. Considering the fact that a significant share of technologies used for this very purpose is based on the application of various plugging materials, knowledge of theoretical and practical aspects of plugging systems application is one of the key factors increasing the field operation efficiency. Unfortunately, today the majority of educational institutions of higher education do not provide well cementing as a separate course. They cover only some aspects of the process in the context of various related subjects. Such an approach eventually leads to significant gaps in the integrity of the ideas of young professionals, and graduate and undergraduate students about cementing operations. In this book, the authors tried to cover all major aspects of oil and gas well cementing technology – from the initial phase of Portland cement production to the practical calculations carried out during complex cementing operations. However, the way the material is presented in the book is based on the logical sequence of cementing operations, which significantly simplifies the reader's perception, despite the wide range of presented information.

The first chapter of the book introduces the reader to the basic concepts used by experts in the plugging process, such as the concept of well design, goals, objectives, and basic methods of well cementing. Chapters 2–4 are devoted to production, properties, types, and classification of plugging materials. The use of cementitious materials in human life has its roots in antiquity: clay and lime have long been used by humankind as a binding material in masonry, and ancient

Egyptians used unrefined lime in the construction of the pyramids. However, with the increase in the industrial production of cement, the range of cementitious materials increased significantly and there was a need for their standardization and classification. These chapters introduce the reader to the modern production process of cementitious materials, chemical composition, and additives used to regulate their properties. The two most common classifications of Portland cement according to API and QOST standards are also given.

The fifth chapter of the book is devoted to surface and downhole equipment used in cementing operations. It contains schematic drawings of almost all types of modern cementing equipment, describing their operating principles and mechanisms. The material allows the reader to understand the main tasks they perform and to become familiar with the technical limitations of their application, without being immersed in complex engineering descriptions of the devices and mechanisms.

Chapters 6–7 provide a detailed description of primary and secondary well cementing. By reading these chapters, the reader will get a sufficiently accurate idea of the types of modern cementing operations, tasks they perform, and factors the engineer is guided by when selecting a cement slurry formulation or type of cementing operations on a well.

Chapter 8 discusses the most commonly used methods for evaluating the quality of casing cementing. The theoretical and practical basics of logging are described in detail, and the advantages and disadvantages of the main methods are listed.

The ninth chapter is devoted to perhaps one of the least covered areas of well cementing, namely laboratory testing methods for cementitious materials. The chapter describes experiments to determine cement slurry properties, outlines extensive theoretical and practical material for a better understanding of the goals and objectives of the experiments conducted, and the role of the parameters to be determined in the casing string cementing process.

The tenth chapter is the final chapter of the book and is of practical application. It explains in detail, with examples, all the basic typical calculations involved in planning or carrying out cementing operations.

The book is intended for researchers, petroleum engineers, postgraduates, students, and specialists in relevant fields.

The authors are grateful to Prof. E.M. Suleimanov for valuable discussion on theoretical aspects of cementing efficiency evaluation, as well as to Prof. N.M. Gadzhiev for valuable discussions on practical aspects of oil and gas well cementing.

Introduction

Well cementing is largely integrated into the well construction process, being one of the two most frequent operations involved. The process should be considered solely in terms of the processes that precede and follow it. Such an approach provides the most complete picture of the purpose and objectives of well cementing while making the learning material much easier to grasp.

In well construction, the drilling phase can be simplified into the following major cycles (Figure 1):

- Drilling
- Casing and cementing the well
- Drilling into the reservoir and testing for oil and gas flow

The first two cycles alternate (i.e. drilling and casing), and once the pay zone has been reached, the process is completed by penetrating the reservoir and testing for oil and gas flow [1].

The number of such repeated cycles is individual for each well and depends on specific geological conditions.

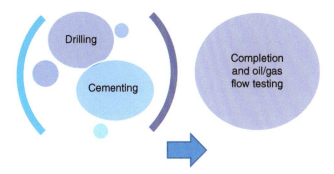

Figure 1 Simplified scheme of the well construction technological cycle.

1

Theoretical and Practical Aspects of Well Cementing

1.1 Oil Well, Its Elements, and Construction

What is oil and gas drilling? According to the classic definition: "Well drilling is the process of making a directional cylindrical rock hole in the earth, the diameter of which is tiny in comparison to its length, without human access to the bottom hole." (A hole drilled into the earth for the purpose of exploring for or extracting oil, gas, or other hydrocarbon substances.) A **well** is a cylindrical rock excavation that has a diameter many times smaller than its length (Figure 1.1).

The point where the drilling starts at the surface is called the **wellhead**, at the opposite end of the excavation is the **bottom hole**, and the entire borehole volume between these two points is called the **borehole**. An imaginary line drawn through the centers of the cross sections of the borehole is called the **borehole axis**.

Drilling begins with running the **drill string**. The drill string is an assembly of pipes connected by interlocks and designed to supply hydraulic and mechanical energy to the bit and create an axial load on it, as well as to control the path of the borehole being drilled. The **drill bit** is the main element of the drilling tool for the mechanical destruction of rock on the bottom hole in the process of penetrating (drilling) the borehole (Figure 1.2).

In the traditional drilling method, the drill bit is connected to the bottom end of the bottom hole assembly (BHA), i.e. the first weighted drill pipe (**Heavy Weight Drill Pipe – HWDP**), while the top end of the **BHA** is connected to the lead drill pipe (**Kelly**). The kelly, a square or hexagonal pipe, which is located at the top of the drill string, serves as the connection between the drill string and the drill rig to ensure the effective interaction between them (Figure 1.3). The drill string then begins to rotate and the drilling process begins. It should be noted that when a **Top Drive** is used, the top end of the BHA is connected to the power swivel.

Oil and Gas Well Cementing for Engineers, First Edition. Baghir A. Suleimanov, Elchin F. Veliyev, and Azizagha A. Aliyev.
© 2023 John Wiley & Sons Ltd. Published 2023 by John Wiley & Sons Ltd.

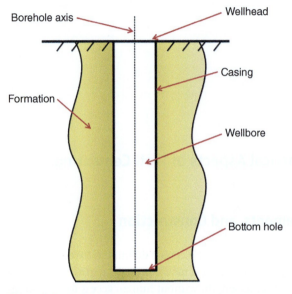

Figure 1.1 Well construction elements.

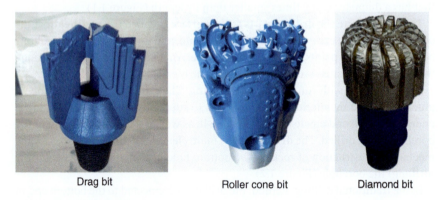

Figure 1.2 Drill bit types.

Once the drill string is deepened by the length of the kelly, the rotation is stopped, the drill string is raised, a new pipe is added, and the drill string is run back into the borehole and drilling is continued. This operation is repeated until the required depth is reached or the drill bit is replaced (Figure 1.4).

There are two main types of drilling based on the area where the rock destruction tool (bit) affects the bottom hole:

1) **Drilling along the whole face area.** Continuous bottom hole drilling, in this type of drilling, the whole rock in the borehole is destroyed by the bit (Figure 1.5a).
2) **Drilling along the peripheral part of the face (annular face).** Core drilling, this type of drilling preserves the inner part of the rock (Figure 1.5b).

Depending on the curvature of the borehole, wells are subdivided (Figure 1.6) into the following types:

- Vertical (Figure 1.6 [1])
- Slant (Figure 1.6 [2])
- Inclined (Figure 1.6 [3])
- Horizontal (Figure 1.6 [4])
- S-type (Figure 1.6 [5])

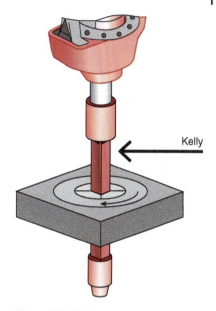

Figure 1.3 Kelly.

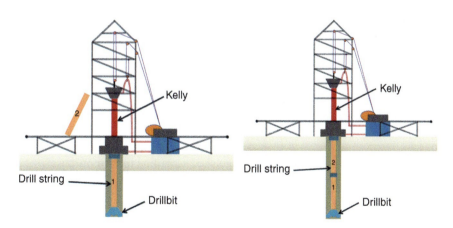

Figure 1.4 Schematic representation of the drilling operation.

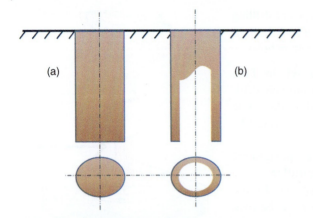

Figure 1.5 Schematic representation of drilling along the whole face area (a) and drilling along the peripheral part of the face – annular face (b).

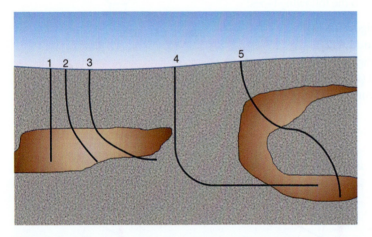

Figure 1.6 Types of boreholes according to borehole curvature.

The following types of wells are distinguished according to their purpose:

1) **Production wells.** Those drilled directly for production by the well (i.e. oil, gas, and condensate).
2) **Injection wells.** Drilled to maintain reservoir pressure and enhance oil recovery by injecting different displacing agents into the reservoir (e.g. water, gas, and polymer solutions).

3) **Exploration wells.** Wildcat and appraisal wells are collectively referred to as exploration wells:
    a) **Wildcat well.** Drilling an exploration well to determine the existence of petroleum in a probable hydrocarbon deposit.
    b) **Appraisal well.** Drilled, as a rule, after the wildcat well to assess the hydrocarbon reserves, collect geophysical information, and outline the reservoir.
4) **Special wells.** Observation, parametric, and stratigraphic – drilled to study the dynamics of reservoir properties, pressure, and degree of depletion in certain reservoir sections, as well as to ensure the in situ burning. The goals and objectives of this group of wells vary considerably from field to field and are determined by the course of the development process and the individual peculiarities of a reservoir.
5) **Structural exploration wells.** These wells include all the wells drilled in the exploration area before the commercial flow of oil or gas is achieved. These wells are generally of small diameter and depth to reduce the cost of the drilling process.

1.2 Objectives of Well Cementing

The geological conditions, which change with depth, impose certain limitations on the drilling process, and it is not possible to reach a productive formation in a single drilling trip.

A **drilling trip** is a set of basic and auxiliary activities to deepen a well with a single rock-drilling tool (bit), starting from the preparation of the drilling tool to be run into the well and finishing work after it has been lifted.

The well trajectory, divided into sections composed of formations with similar geological characteristics and therefore compatible in terms of drilling conditions, is called **drilling intervals**.

Each drilling interval has different requirements for both the drilling process and the technology, i.e. the intervals have different drilling conditions. In other words, it is not possible to drill the underlying interval without complications in the overlying interval drilled, unless the latter is secured with casing and cementing. Let us consider, in a simplified way, the process of drilling intervals selection by the example of conventional well "A" (Figure 1.7). Initially, based on geological conditions **pore pressure gradient (PPG) graph** is constructed, which illustrates the change along the depth of the well: hydraulic fracturing pressure, formation pressure, and hydrostatic pressure of the drilling fluid column. The PPG graph, taking into account geological complications, makes it possible to identify drilling intervals in the well trajectory and identify the need for intermediate casing strings, their number, and depth of penetration. In this example, the PPG chart clearly identifies three zones that are incompatible in terms of drilling conditions.

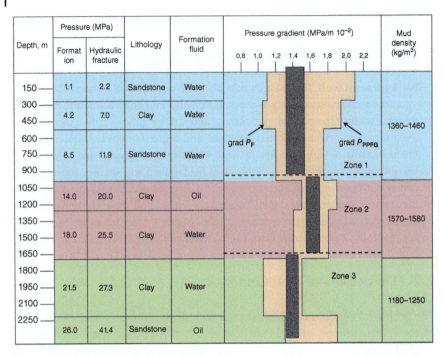

Figure 1.7 Pore pressure gradient graph.

Thus, drilling "Zone 2" with the same mud as "Zone 1" will inevitably result in fluid influx because, for "Zone 2," the hydrostatic pressure of the drilling fluid column will already be below the reservoir pressure. Drilling "Zone 3" with the same density as "Zone 2," on the other hand, will result in hydraulic fracturing. In order to meet the compatibility conditions, these zones must be bridged with casing, increasing the number of intermediate casing strings in the well design to two.

It should be noted that even if the well would consist of a single drilling interval (which of course is not realistic), an uncased wellbore (i.e. without casing) would be a very unstable structure with a constant risk of rock debris or even collapse into it.

As mentioned earlier, the borehole is secured by running special pipes, known as casing strings, into the borehole. The space between the casing and the wellbore wall is filled with a plugging material. The most widely used plugging material is mortar on the basis of **Portland cement**, hence the name of this process: **well cementing**. The main functions of cementing include:

- Isolation of layers, i.e. prevention of fluid transfer from one layer to another. For example, isolating freshwater formations from mixing with highly saline formation water.
- A cement sheath around the casing contributes to preventing gas and water seepage into the well.

1.2 Objectives of Well Cementing | 7

- Protection against corrosive effects of formation fluids on the casing.
- Protection of casing against drilling stresses.
- Sealing of absorption zones.
- Cementing is also applied when abandoning depleted wells or isolating depleted layers.

The alternation of drilling and cementing processes results in a stable underground structure called a well construction. Thus, a well construction is a set of data reflecting the following information:

- Casing lengths and diameters
- Wellbore diameters corresponding to the casing used to secure the interval
- Cementing intervals
- Completion method and payzone interval

Casing is usually classified according to its purpose, so there are four main types of casing (Figure 1.8).

Let us take a closer look at each casing type.

Conductor casing – This type of casing is designed to reinforce the wellhead against collapse and destruction as a result of the drilling process and to allow circulation of drilling mud. A conductor is usually comprised of a single casing, less frequently a single pipe, and the pipes used as a conductor are not pressurized. The running depth of a conductor could be as much as 100 m (extended direction), but in practice, it rarely exceeds 10 m; the diameter of the casing ranges from 245 to 1250 mm. The conductor is secured either by concreting or simply by driving the casing into the ground (piling method).

The ***surface casing*** is the subsequent casing string running to a depth of 100–600 m, with an average casing diameter of 177–508 mm. The main purpose of running the surface casing is to prevent contamination of the freshwater horizons in the upper horizon of the rock section, as well as to install wellhead equipment (i.e. casing head and blowout control equipment). Unlike the conductor, surface casing and the cement ring behind it are compressed.

It should be noted that the conductor and surface casings are fixed parts of all well designs.

Intermediate casing – The main purpose of running intermediate casing is to separate different drilling intervals. As there may be several drilling intervals, depending on geological conditions, the number of intermediate casing strings run is often more than two and varies from well to well.

The following types of intermediate casing are distinguished:

- **Continuous intermediate casing.** The entire annular space from the bottom hole to the wellhead is cemented.
- **Liner.** Used for casing the non-cased section of the wellbore. In this case, the casing itself is installed by overlapping a part of the previous casing string and

1 Theoretical and Practical Aspects of Well Cementing

(a)

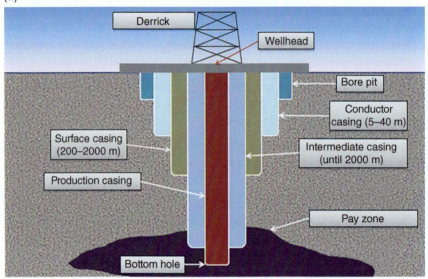

(b)

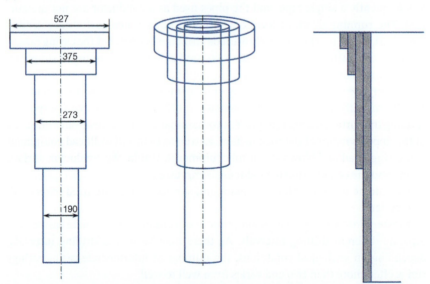

Figure 1.8 Different types of schematic representation of casing. (a) Schematics of well construction in relation to payzone; (b) different types of well construction drawings.

does not extend to the surface. Running liner is often economically advantageous as it allows for a significant simplification of well construction and reduction of metal and insulation material consumption.
- **Tie-back liner** is a special type of casing that is not connected to other casing strings included in the well design and serves to isolate the selected interval.

The production casing is the final casing string run to isolate the reservoir and bring the well product to the surface. In practice, it is not uncommon for the last intermediate casing to perform the role of production casing, and the production casing itself is used not only to produce oil or gas but also to inject fluids into the reservoir.

Well design development is based on the following main geological and technical-economic factors:

- geological features of rock occurrence, their physical and mechanical characteristics, the presence of fluid-bearing horizons, reservoir temperatures and pressures, and the fracturing pressure of the rocks being penetrated;
- the purpose and aim of drilling the well;
- the intended method of completing the well;
- the method of drilling the well;
- the level of organization, technique, drilling technology, and geological knowledge of the drilling area;
- level of qualification of drilling team and organization of logistics; and
- methods and techniques of well completion, operation, and workover.

1.3 Primary Cementing

There is no universal technology of cementing because this process depends on many factors, such as geological conditions, well design and its current condition, cementing team logistical support, etc. Therefore, the selection of cementing technology is performed individually for each particular well, but nevertheless, there are the following general requirements, which should be met by the chosen technology of cementing:

- Provision of safe anchoring of the whole borehole section.
- Preventing drilling mud contamination with cement slurry.
- Complete displacement of drilling mud from the cemented interval.
- Obtaining a quality cement ring in the annulus.

In practice, the following well cementing techniques are the most commonly used:

- Single-stage cementing with two plugs
- Two-stage cementing

- Basket cementing
- Liner cementing
- Reverse cementing
- Cement plug installation

Let us take a short description of each technology separately. (*Note*: Each technology will be described in more detail in Chapter 6.)

1.3.1 Single-Stage Cementing with Two Plugs

Well cementing operation in single stage using two cementing plugs was first carried out in Baku (Azerbaijan) in 1905 by engineer Bogushevsky. Simplified this technology looks as follows (Figure 1.9):

1) After lowering the casing, a cementing head is installed to connect it tightly to the injection lines of the cementing units. The wellbore at this stage is filled with drilling fluid. Before pumping the cement slurry, the bottom plug is released from the cementing head, which serves as a kind of barrier between the drilling fluid in the borehole and the cementing slurry. It should be noted that in addition to cementing plug serving as a mechanical barrier between fluids in the wellbore, not infrequently in order to create a natural barrier, spacer fluids are also used.
2) After pumping a calculated amount of cement slurry into the well, the upper plug is released, which, in turn, is forced down the wellbore by drilling fluid. Thus, the cement slurry column in the wellbore is actually limited on both sides by cementing plugs.

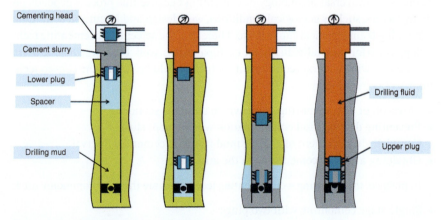

Figure 1.9 Single-stage cementing with two plugs.

3) Unlike the upper plug, which is a solid plug, the lower plug has a rubber diaphragm, and upon reaching the stopper ring with increasing wellhead pressure, the diaphragm ruptures and the cement slurry rushes into the annulus.
4) When the top plug reaches the stopper ring (*Note*: The top plug actually sits on the bottom plug), because of the solid design the diaphragm is not destroyed and the pressure at the cementing head increases sharply, which serves as a signal to stop pumping the flushing fluid. The well is sealed and left for the time required for setting of cementing slurry.

As a rule, some amount of cement slurry remains under the plug in the casing, forming a so-called **shoe track** of 10–25 m height. If further drilling is required, the upper cement plug and shoe track are drilled out.

1.3.2 Two-Stage (Two-Cycle) Cementing

This technology divides the cemented section of the wellbore into two intervals (upper and lower), and each interval is cemented separately (Figure 1.10).

Two-stage cementing has the following advantages over single-stage cementing technology:

- Significantly lower hydrostatic pressure of the cement slurry column on the formation, thus avoiding a number of complications when cementing deep wells.
- It allows avoiding a considerable increase in injection pressure due to a decrease in cement slurry lifting height in the annulus.

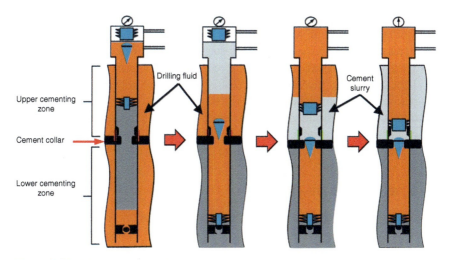

Figure 1.10 Two-stage cementing.

- The cement slurry used for the upper interval is much less susceptible to high temperatures, which simplifies the cement slurry formulation for this interval while reducing the consumption of the used reagents.

Two-stage cementing technology can be simplified into the following stages:

- A cementing sleeve is installed at the interface between the upper and lower cementing intervals. The process of preparing the well for work is absolutely identical to the previous method. The volume of cementing slurry necessary for setting the lower interval is pumped into the borehole. At that, as well as in the first method, drilling and cementing slurry are separated with cementing plugs. Both plugs freely pass through the cementing (filling) sleeve. The cement slurry is displaced into annular space by the casing volume, which corresponds to the casing volume in the lower cementing interval.
  Note. In practice, it is often the case that the bottom plug is not used, and the entire process is carried out using only three plugs, two for the upper and one for the lower cementing interval.
- After pumping the calculated volume of the displacement fluid, the bottom plug of the upper cementing interval is released. It does not pass through the cementing sleeve freely, but sits on a special sleeve, sliding it down. As a result, the sleeve opens the holes in the annulus of the upper cementing interval, accompanied by a sharp drop in injection pressure at the wellhead.
- Then there are two scenarios for the development of the process:
  1) The bottom plug of the upper cementing interval is squeezed with cement slurry. This method is called the two-stage method of continuous cementing.
  2) The bottom plug of the upper cementing interval is squeezed with drilling mud, and the cement slurry is pumped after a certain time (usually necessary for the formation of a strong cement ring in the lower cementing interval). This method is referred to as the two-stage cementing method with a breakout.
- After pumping a calculated volume of cement slurry to secure the upper interval, the upper plug (the last one) is released and pressed with drilling fluid until it reaches the cement collar. During the setting of the plug in the cement collar, the sleeve shifts, blocking access holes to the annulus and causing a sharp increase in pressure, which signals the completion of the process. The well is then left to rest for the time necessary to form a strong cement ring.

1.3.3 Basket Cementing

The current method of cementing is widely used in case of low formation pressure in productive formation or for preventing cementing fluid contamination of perforated interval of wellbore (i.e. screen interval). The technology consists of installing a cement collar at the level of the lower cementing interval and a basket made of

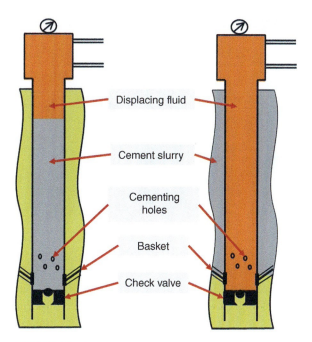

Figure 1.11 Basket cementing.

high-strength metal and features overlapping fins on the outside, hence the name of the current method of cementing (Figure 1.11). This design prevents the cement slurry from moving down the wellbore, directing the flow in one direction only: up. A valve is installed inside the casing itself, which performs an identical function, that is, it does not allow cement slurry to penetrate to the bottom of the casing.

1.3.4 Liner Cementing

In this method of cementing, the casing is run in sections using drill pipes, with casing sections connected to drill pipes with a left-hand transfer tube (i.e. a left-hand threaded transfer tube) (Figure 1.12). Another feature of this technology is the use of a two-part separating plug:

- **The female plug** is of the same diameter as the inner diameter of the casing section to be cemented. The borehole plug is mounted at the junction of the casing and drillpipe.
- **Male plug** – with a diameter smaller than the inner diameter of the drill pipe.

The male plug, installed in the cementing head, is released after the calculated volume of cementing solution is pumped and pressed with drilling fluid until it

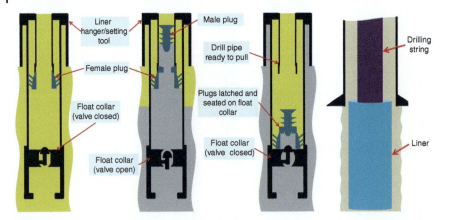

Figure 1.12 Liner cementing.

reaches the female plug. The male plug gets stuck in the female plug, causing an increase in pressure, as a result of which the pins holding the female plug are cut off and both plugs move down the wellbore to the stopper ring as a single unit. As the plugs reach the stopper, the pressure rises sharply, signaling the end of the operation and the presence of the pumped cement slurry in the annulus. Then the drill pipe is unscrewed from the left sub and flushed from the cement slurry residue and removed from the borehole.

1.3.5 Reverse Cementing

Reverse cementing is probably the least widespread cementing technology in practice. The essence of the technology consists in pumping the cement slurry directly into the annular space with the displacement of the drilling fluid along the inner diameter of the casing to the surface (Figure 1.13). The reason for the absence of wide industrial application is the fact that from a technical point of view, it is very difficult to determine the moment when cement slurry reaches the bottom part of the casing and as a result to provide the formation of quality cement ring in this interval.

1.3.6 Cementing Plugs

A cement plug is a cement column in a wellbore, usually up to several tens of meters high, to isolate a certain interval of the wellbore (Figure 1.14). The reasons for installing cement plugs may vary from switching to operating these productive formations to creating an artificial downhole face when sidetracking.

1.3 Primary Cementing

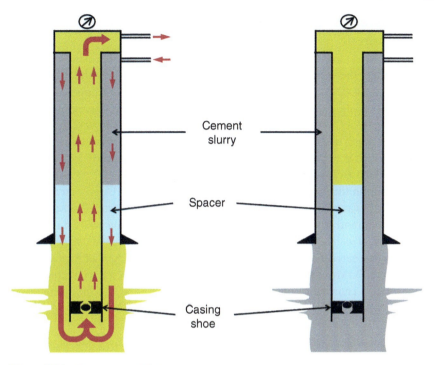

Figure 1.13 Reverse cementing.

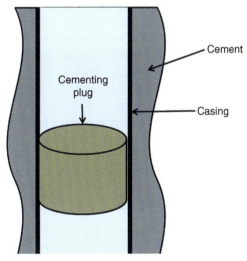

Figure 1.14 Cementing plugs.

1.4 History of Oil Well Cementing Technology Development

In 1844, under the leadership of State Councilor Vasily Semyonov and the director of the Baku oil fields, Major Alexeyev of the Corps of Mining Engineers, more than 10 years before the drilling of Edwin Drake's famous well in Pennsylvania, drilling began in the Bibi-Eybat field (Azerbaijan). In 1847, the first exploratory well with a depth of 21 m was drilled there. The first modern oil well in the world was also drilled here between 1847 and 1848. The first oil was produced by percussion with the use of wooden rods on 14 July 1848.

However, the first use of cement in well construction (to block water inflows into the well) was made by Frank Hill in 1903 in the United States almost 60 years later. Cement mortar was mixed manually and poured into the well with a rod.

Operation of well cementing in one stage with the application of two cementing plugs was first carried out in Baku (Azerbaijan) in 1905 by engineer Bogushevsky. In 1910, A.A. Perkins also carried out one-stage cementing of wells with the application of two cementing plugs. In 1918, he also founded the first cementing service company in Los Angeles, United States.

In 1919, Earl Halliburton established a well cementing service in Texas. Over time, the number of cementing service companies increased dramatically. At the same time, as drilling depths increased and cementing conditions became more complex, it became necessary to control the setting time of the cement slurry. Oatman [2] was one of the first pioneers in this field, having applied calcium chloride to reduce the setting time and setting strength of cementing stone. The development of setting time retarders was a technically more difficult task and required an appropriate material and technical base. It became necessary to establish highly specialized research centers, and in 1929, Halliburton established its first cement laboratory and in 1930, together with other companies (Humbly Oil and Refining Co., Standard Oil Co. of California) founded a research center dedicated to well cementing. Within seven years, in 1937, American Petroleum Institute (API) founded the Committee for Cement Research, which today includes dozens of laboratories and research centers. In the next 30–40 years, there is a rapid development of both cementing technology and the production of cement and cement additives. In 1940, two types of cement and three types of cement additives were used. By 1975, according to API specification, there were four types of cement and 44 types of additives. At this stage, there is an urgent need for standardization of testing and research processes for cement slurries. In 1952, the first API standard in this area is published API32. Since that period, the API publishes two booklets each year:

- **API spec 10A** is actually a classification of cement according to API standards.
- **API RP 10B** is a practical recommendation for testing cement slurries.

1.4 History of Oil Well Cementing Technology Development

To date, each of these standards has undergone more than a dozen revisions.

There was no less rapid development of material and technical means used in the process of well cementing. For example, the first centralizers for providing an even gap between the casing string and wellbore were patented and used by H.R. Irvin in 1930, but their mass application started only in 1946. In 1934, Schlumberger suggested and patented the method of cementing quality evaluation based on thermal logging data. In 1948, G.S. Howard and D.B. Clark published a number of scientific papers on fluid dynamics in cementing, and the results of these works are still used today.

Production of cementing equipment has also been actively developing. In 1920, Halliburton introduced into field practice the preparation of cement slurry with the use of hydro-vacuum mixer, as well as patented designs of a number of cementing pumps. In 1962, Wugon Jackson designed and manufactured the Pacemaker triplex pump, considered the best in its class today.

2

Composition and Classification of Portland Cement

2.1 Chemical Composition

The chemical composition of Portland cement consists of 95% of four chemical compounds, and the remaining 5% are compounds such as gypsum, sodium and potassium sulfates, and magnesium (Table 2.1). In the production of construction cements, minor admixtures are usually not taken into account, but in the production of oil-well cement, they are very important due to their influence on such properties as the rate of hydration of cement slurry and sulfate resistance of cement stone.

Tricalcium silicate (C_3S/alite). It is formed by the reaction of lime and silica and has the highest mass content in the cement compound. When mixed with water, the hydration reaction produces a tricalcium silicate hydrate gel and lime, which plays a role as a filler in the cement stone, usually without having a negative effect on it. However, it should be borne in mind that this component is less resistant to acidic fluids. With increasing mass content of tricalcium silicate in cement compound, the strength of cement stone also increases. So, if in standard cements the alite content is from 40% to 45%, in high-strength cements, it reaches 65%. C_3S — is the main chemical component in the Portland cement responsible for increasing the strength of the cement stone, especially in the early stages of hardening.

Dicalcium silicate (C_2S/belite) is also formed by the reaction of calcium oxides (CaO) and silicon (SiO_2). However, unlike alite, the hydration reaction of dicalcium silicate is slower, respectively, and has little effect on the initial setting of the cement. This component provides a slight gradual increase in the strength of the cement stone over a long period of time.

Oil and Gas Well Cementing for Engineers, First Edition. Baghir A. Suleimanov,
Elchin F. Veliyev, and Azizagha A. Aliyev.
© 2023 John Wiley & Sons Ltd. Published 2023 by John Wiley & Sons Ltd.

2 Composition and Classification of Portland Cement

Table 2.1 Chemical composition of Portland cement.

Component	Chemical formula	Definition	Concentration (mass %)
Tricalcium silicate (alite)	$3CaO * SiO_2$	C_3S	36–65
Dicalcium silicate (belite)	$2CaO*SiO_2$	C_2S	15–40
Tetracalcium aluminoferrite (braunmillerite)	$4CaO*Al_2O_3*Fe_2O_3$	C_4AF	7–17
Tricalcium aluminate	$3CaO. Al_2O_3$	C3A	5–16

Tricalcium aluminate (C_3A) is formed by the reaction of calcium oxides (CaO) and aluminum (Al_2O_3). C_3A has no significant effect on the final strength of hardened cement, but actively reacting with cement additives, greatly accelerates the hydration reaction (i.e. earlier development of cement stone strength). Gypsum is often used to regulate the setting time of C_3A.

Tetracalcium alumoferrite (C_4AF/braunmillerite). Under comparable conditions, the hydration products are similar in many respects to those produced by the hydration of C_3A, although the reaction rates differ. C_4AF has the highest rate of hydration of all four clinker phases. However, the reactivity of this phase usually decreases with increasing iron (Fe) content.

The mixture of the four reagents listed here is called clinker, but under certain technical conditions, other materials may also be added to the grinding process. In essence, the cement stone formation process is related to the hydration reaction of clinker. Since alite has the highest mass content in the cement compound, it determines most of the basic properties of cement. The cement hardening process itself is gradual, lasting up to several days.

This process can be described in a simplified way as follows: Initially, there is a so-called "direct contact of reactants" stage, i.e. there are no barriers to the interaction of water and clinker (Figure 2.1). As the hydration reaction proceeds, two chemical compounds – calcium hydrosilicate (C–S–H, chem. $3CaO. 2SiO_2. 3H_2O$) and calcium hydroxide (lime – $Ca(OH)_2$) – are formed.

$$6CaO * 2SiO_2 + 6H_2O \rightarrow 3CaO * 2SiO_2 * 3H_2O + 3Ca(OH)_2 + 114\,kC/mol$$

At this stage, the formed calcium hydrosilicate (C–S–H) prevents direct contact with water and clinker, acting as a kind of barrier, significantly slowing down the hydration reaction, but not stopping it. The hydration process is complete only in the absence of unbound water in the pores. Thus, the hardening reactions take place in both phases:

- in the liquid phase, the so-called "fast" reactions,
- in the solid phase – topochemical (slow) reactions.

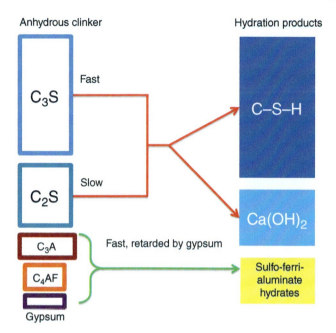

Figure 2.1 Schematic representation of the Portland cement hydration process.

Of course, this description of the hardening process of cement stone is simplified and only serves to illustrate the complexity of the phenomena occurring in the process of the formation of cement stone.

It is more convenient and clear to examine the process of cement hydration according to the data of calorimetric measurements. Based on the results of these measurements, the following five stages in the hydration process are distinguished (Figure 2.2).

Stage I is the most rapid stage characterized by the formation of calcium hydrosilicate (C–S–H) and lime (Ca(OH)$_2$), with the release of large amounts of heat. Lime crystallizes when it reaches saturation concentration.

Stage II – the calcium hydrosilicate (C–S–H) gel formed on the alite surface creates a barrier to water, slowing down the hydration reaction. At this stage, the heat release is significantly reduced.

Stage III – occurs during the period from four to eight hours after the beginning of stage I and is characterized by a breakdown of the structure of calcium hydrosilicate gel (C–S–H), and the continuation of the hydration reaction of alite due to the penetration of new portions of water. This period is characterized by the release of heat and an increase in the strength of the cement stone.

Stage IV – as a rule, stages 3 and 4 are combined in one segment, as, in fact, the difference between them is weakly expressed and appears only in the formation of

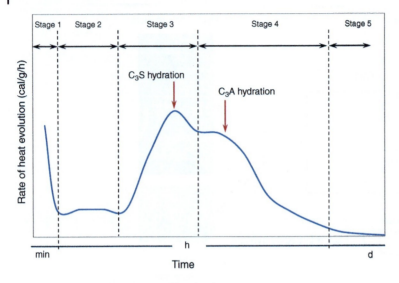

Figure 2.2 Hydration process of Portland cement.

a new small peak on the calorimetric curve characterizing the hydration of belite. In other words, at this stage, the hydration reaction proceeds mainly with the participation of tricalcium aluminate (C_3A).

Stage V – for most cements comes, in general, 24 hours after the beginning of the stage I. This period is characterized by a significant slowing of the hydration reaction and negligible heat release.

2.2 Portland Cement Manufacturing

Clay and lime have long been used by people as a binding material in masonry. As far back as the Roman Empire, pozzolans were widely used as a cementitious material in construction, and the ancient Egyptians used unrefined lime in the construction of the pyramids. The production of modern Portland cement was associated with the beginning of the production of quicklime. In the process, it was discovered that the presence of clay inclusions in the raw materials had a positive effect on the quality of the final product. Taking advantage of this discovery, in 1796, James Parker created a material he called "Roman concrete"; one of the first cements to be water resistant after hardening.

Joseph Aspdin, owner of a cement factory in Wakefield, England, patented in 1824 a compound he called "Portland cement," for its similarity in color to limestone mined on the island of Portland off the south coast of England.

Today, of course, the process of cement production has changed and differs from the production technology that Aspdin used. In principle, however, cement production is still a process of grinding clinker and gypsum in certain proportions, and the clinker is obtained by burning limestone and clay.

The burning is done in special rotary furnaces, a complex structure consisting of several zones, which can be conditionally divided as follows:

- Heating zone
  - At temperatures of +200 °C to +650 °C, the clay dehydrates and burns out the organic components
  - At temperatures of +600 °C to +1000 °C, the aluminum silicates decompose
- Decarbonization zone.
  - At temperatures between +900 °C and +1200 °C, the CO_2 content of the limestone decreases. At the same time, new chemical compounds are formed, i.e. processes of so-called solid phase synthesis take place (CaO–Al_2O_3);
- Exothermic reaction zone
  - At temperatures between +1200 °C and +1350 °C, the end of solid-phase synthesis takes place and the formation of clinker components such as belite, cellite, and felite.
- Sintering zone
- At temperatures between +1300 °C and +1480 °C, a partial melting process takes place with the formation of the last clinker component alite (C_3S).
- Cooling zone
  - At temperatures from +1300 °C to +1000 °C, the mixture gradually cools down.

Depending on the method of obtaining the mixture for burning, the following three methods of Portland cement production are distinguished:

Dry. This technology is not uncommonly called the "quick" method of Portland cement production due to the combination of some technological stages of the process. Drying and grinding of clinker components take place simultaneously in special mills, in which hot gas mixtures are fed (Figure 2.3).

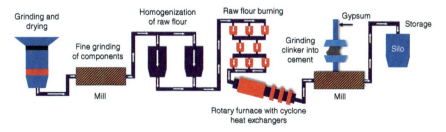

Figure 2.3 The "dry" method of Portland cement production.

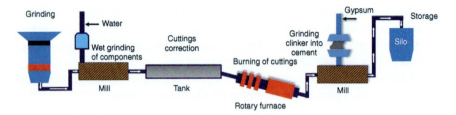

Figure 2.4 "Wet" method of Portland cement production.

Wet. In this method of production, the initial raw material is crushed to obtain certain fractions, so, for example, the average particle diameter of limestone should not exceed 10 mm. The clay is stored in separate tanks, soaking in water until obtaining 70% moisture. Upon reaching the necessary conditions clay and limestone are mixed in mills (Figure 2.4).

Semi-dry (combined). This technology of cement production combines the technological processes of the two previous methods. In fact, it boils down to reducing the moisture of charge material to 14–18%, that is not only a complete dehydration of raw materials as in the "dry method" of production but also not so high degree of moisture as in the "wet method" of production. The further process is identical to the previous two, the obtained raw materials are mixed and burned in special mills.

Clinker obtained by one of the three methods described earlier is a bulk material with a fairly wide range of particle size distribution from a few microns to several tens of millimeters. The clinker is cooled and, after a certain period of storage, is crushed and mixed with gypsum. The mass content of gypsum ranges from 1.5% to 3% depending on the required setting time of the cement produced. The resulting material is packaged in special containers or stored in hoppers – this powdered mixture is called **Portland cement**.

2.3 API (American Petroleum Institute) Classification of Portland Cement

As the volume of industrial cement production increased, the range of cement types produced also increased significantly. It became necessary to standardize and classify cement types. For this purpose, in 1937, American Petroleum Institute (API) created subcommittee number 10. Subcommittee 10 still exists today and works with ISO (International Organization for Standardization) to set current specifications for cement systems. It should be noted that there are other classifications of Portland cement which are not widely used. Thus, in the United States, they quite often use ASTM classification (American Society for Testing and

2.3 API (American Petroleum Institute) Classification of Portland Cement

Table 2.2 Main cement classes according to API classification.

API class	Slurry density, kg/m^3 (ft/gal)	Depth (m)	Temperature, (°C)	Water, % by weight of cement	Free water, max %
A	1869 (15.6)	0–1800	26–76	46	—
B	1869 (15.6)	0–1800	26–76	46	—
C	1773 (14.8)	0–1800	26–76	56	—
D	1965 (16.4)	1800–3000	76–110	38	—
E	1965 (16.4)	3000–4200	110–143	38	—
F	1941 (16.2)	3000–4900	110–160	38	—
G	1893 (15.8)	0–2400	26–93	44	5.9%
H	1965 (16.4)	0–2400	26–93	38	5.9%

Materials), which distinguishes five basic types of cement. API distinguishes eight basic types of cement, such difference is due to the fact that according to ASTM classification, cement is mainly produced for civil construction (Table 2.2). Of course, due to the fact that the same physical and chemical properties of cement are used for classification, there is a correlation between different classes of cement in all classifications. The most common and applicable classification, however, remains the API classification.

Of the eight API cement classes, five are most commonly used: A, B, C, G, and H. Classes D, E, and F are hardly ever used in the United States, Canada, and very rarely in other countries of the world. About 80% of cements used in oil well cementing are of classes G and H, with cement of class G prevailing considerably. It should be noted that API also subdivides Portland cement into grades according to sulfate resistance:

- Ordinary cement (O),
- Moderate sulfate-resistant cement (MSR),
- High sulfate-resistant cement (HSR).

Clinker composition for the main cement classes according to the API classification is given in Table 2.3. The following basic patterns in changing clinker composition to achieve certain physical and chemical properties should be noted:

- Increase of early strength values of cement stone, achieved by increasing the content of C_3S
- Cement mortar hardening time depends on the C_3S and C_3A concentration
- The reduction of heat release during hydration can be achieved by decreasing the C_3S and C_3A concentration
- Cement sulfate resistance increases with decreasing C_3A concentration.

Table 2.3 Clinker composition for the main cement classes according to API classification.

API cement class	Composition (%)			
	C_2S	C_3S	C_3A	C_4AF
A	53	24	>8	8
B	47	32	<5	12
C	58	16	8	8
D and E	26	54	2	12
G and H	50	30	5	12

Let us look at a brief description of each cement class.

Class A cement

Description. Only one grade is produced according to the degree of sulfate resistance – standard (O), the concentration of additives used is regulated by ASTM C 465, according to ASTM classification, it corresponds to ASTM C 150, type I.

Application. This class of Portland cement is used in uncomplicated cementing conditions at a depth of up to 1800 m or at a downhole temperature not exceeding 76 °C. The specification of Class A cement is more in line with cements used in the construction industry. When cementing wells, as a rule, this type of cement is used for surface or conductor casing cementing, taking into account that the depth and temperature of cementing interval correspond to the aforementioned criteria. It should be noted that due to high variability in the composition of this cement from different manufacturers, it is critically necessary to conduct thorough laboratory research while selecting a cement slurry composition.

Class B cement

Description. According to sulfate resistance degree there are two types of cement, middle (MSR) and high (HSR), the concentration of additives used is regulated by ASTM C 465, according to ASTM classification it corresponds to ASTM C 150 type II.

Application. This class of Portland cement is designed for application at a depth of up to 1800 m or at a downhole temperature not more than 76 °C in conditions that require sulfate-resistant cement to be used.

Compared to Class A cement, it contains less belite (C_3A) and is more coarsely ground, resulting in a longer thickening time and slower strength development.

Class C cement

Description. According to sulfate resistance degree there are three grades of ordinary (O), middle (MSR), and high (HSR) sulfate resistance, the concentration of additives used is regulated by ASTM C 465, according to ASTM classification, it corresponds to ASTM C 150 type III.

Application. This class of Portland cement is designed for application at a depth of up to 1800 m or downhole temperature not exceeding 76 °C in conditions requiring high initial strength and sulfate resistance of cement used. However, it is necessary to note that in field practice, there are a great number of examples of successful application of the given type of cement at depths more than 3000 m. Initially, low density of cement slurry of $1.7 \, kg/m^3$ (14.8 ppg) significantly increases possible applications of this class of cement. There is quite a wide range of this cement class on the market produced under different brand names, but regardless of the manufacturer, all cements of this class are characterized by the absence of belite (C_3A) in their composition. After all, it is belite (C_3A) that is the most susceptible to sulfate components and a decrease in its content increases the sulfate resistance of cement.

Class D cement

Description. Two grades are produced according to the API degree of sulfate resistance: medium (MSR) and high (HSR) sulfate resistance, the concentration of additives used is regulated by ASTM C 465.

Application. This class of Portland cement is designed for depths of 1800–3000 m or downhole temperatures ranging from 76 to 110 °C. Class D cement belongs to slow setting cements, i.e. cements in which the setting time of the mortar is artificially increased, usually either by excluding quickly hydrating components from the composition or by adding setting retarders. As a setting retarder, starch is mainly used. Today, this class of cement is not widespread and is almost completely replaced by cements of class G and H in field application.

Class E cement

Description. Two grades are produced according to the API degree of sulfate resistance: medium (MSR) and high (HSR) sulfate resistance, the concentration of additives used is regulated by ASTM C 465.

Application. This class of Portland cement is designed for depths of 3000–4200 m or downhole temperatures between 110 and 143 °C (high temperature and pressure conditions). Class E cement, as well as class D cement, is classified as slow setting cement, but as a retarder, primarily lignosulfonates are used rather than starch. At present, this class of cement is not widespread and is almost completely replaced by cements of classes G and H in field application.

Class F cement

Description. Two grades are produced according to the API degree of sulfate resistance: medium (MSR) and high (HSR) sulfate resistance, the concentration of additives used is regulated by ASTM C 465.

Application. This class of Portland cement is designed for depths of 3000–4900 m or downhole temperatures between 110 and 160 °C (high temperature and pressure conditions). It refers to slow setting cements. Lignosulfonates are used as a setting retarder. At present, this class of cement is not widespread and is almost completely replaced by cements of class G and H in field application.

Class G cement

Description. According to the API sulfate resistance grade, two grades are produced: medium (MSR) and high (HSR) sulfate resistance, the concentration of additives used is regulated by ASTM C 465.

Application. This class of Portland cement is designed for depths up to 2400 m or downhole temperatures of up to 93 °C without additives. However, with the addition of additives to control cement slurry properties, it can be used for most wells, both shallow and deep. This cement is produced in accordance with strict requirements for chemical composition.

Class H cement

Description. Two grades are produced according to API sulfate resistance grade: medium (MSR) and high (HSR) sulfate resistance, the concentration of additives used is governed by ASTM C 465.

Application. This class of Portland cement is designed for use in depths up to 2400 m (8000 ft) or downhole temperatures up to 93 °C without additives. However, when additives are added to regulate cement slurry properties, they can be used for most wells, both shallow and deep.

Class G and H cements were developed in the 1960s in the United States. In 1964, cement of class G was patented, which almost immediately became widely used throughout the United States. However, for a number of US fields located in the Gulf of Mexico, it became necessary to use cement mortars with higher density. For this purpose, initially, cement of class H was developed. Both products were produced under strict technical and industrial regulations and differed neither in physicochemical properties nor in composition at different manufacturers. The only difference between these cement classes is the degree of fineness, as cement of class H is milled more coarsely, which reduces the amount of mixing water and results in higher density of slurry. The water-cement ratio in G- and H-grade cements is 0.44 and 0.38, respectively.

It should be noted that it is not uncommon that the water-cement ratio of class H cement is taken as 0.46, which allows for a longer thickening time. A side effect of this practice is a reduction in the initial strength of the cement stone and excessive amounts of unbound water. A more thorough laboratory test is therefore required in this case to ensure that both the cement slurry and the hardened cement have the desired properties.

2.4 GOST (Russian: ГOCT) Classification of Portland Cement

GOST (Russian: ГOCT – **Го**сударственный **Ст**андарт [translate: National Standard]) refers to a set of international technical standards maintained by the Euro-Asian Council for Standardization, Metrology and Certification (EASC), a regional standards organization operating under the auspices of the Commonwealth of Independent States (CIS). (Source: Wikipedia.)

As of today, the GOST 1581-96 is the current state standard regulating the production and classification of cementing materials. Cements are divided into the following five types according to their material composition:

1) **I.** Portland cement without any additive;
2) **I-G.** Portland blast cement without any additive with normalized requirements at a water-cement ratio equal to 0.44;
3) **I-H.** Portland cement without any additive with normalized requirements at water-cement ratio equal to 0.38;
4) **II.** Portland cement with mineral additives;
5) **III.** Portland cement with special additives, regulating density of cement slurry, this type of cement is subdivided into two groups according to density of cement paste:
    - light-weighted (Ob);
    - heavy-weighted (Wt).

In addition, cements according to GOST are classified by temperature of use as follows:

- cements for use in conditions of low and normal temperatures (15–50) °C;
- cements for application in conditions of moderate temperatures (51–100) °C;
- cements for use in conditions of elevated temperatures (101–150) °C.

Just as in the API classification, cement is produced with different degrees of sulfate resistance, so types I, II, and III cements are produced in two grades: ordinary (no requirements for sulfate resistance) and sulfate resistant (CC). In contrast, cements of types I-G and I-H are not produced in ordinary grades. These types of cements are produced either high sulfate resistance (CC-1) or moderate sulfate resistance (CC-2).

2 Composition and Classification of Portland Cement

According to GOST, the name of the cement is written in a certain way and must include the following items:

- letter designation of the cement. For example, PCT (Russian: ПЦТ) – Portland cement;
- cement-type designation;
- designation of sulfate resistance of cement;
- designation of average density for type III cement;
- designation of maximum temperature of use of cement;
- designation of hydrophobization or plasticization of cement – GF (Russian: ГФ) or PL (Russian: ПЛ);
- designations of this standard.

The use of such designations provides sufficiently complete information about the grade, type, and properties of the cement.

For example: PCT III-Ob 5-100-GF GOST 1581-96 is interpreted as follows:

- **PCT.** Portland cement for oil well;
- **III-Ob.** Third type of cement, lightweight
- **5.** Strength grade of oil-well cement (defined by the manufacturer)
- **100.** Application temperature, i.e. cement for moderate temperature applications
- **GF.** Hydrophobized, i.e. resistant to water.
- GOST 1581–96 – standard number

3

Cement Additives

3.1 Introduction

Nowadays Portland cement systems have a very wide range of application, which leads to considerable variation of downhole conditions. It is not rare that wells are drilled in permafrost zones at sub-zero temperatures or, on the contrary, are located in geothermal zones, where temperatures often exceed 300 °C. Formation pressure intervals also vary in a wide range of values, not infrequently exceeding 200 MPa. In addition to harsh environmental conditions, such factors as high mineralization of formation water, highly corrosive fluids used in various well operations, aggressive formation fluids, etc. cause significant complications for the formation of a high-quality and durable cement ring in the annulus. The use of various additives to adjust the properties of Portland cement is what allows to adapt the cement slurries to such a wide and multifactorial environment of plugging operations. In spite of the fact that there are more than 100 different cement additives on the market, there are only eight main categories:

1) **Accelerators**. Chemicals that reduce the setting time of cement slurry.
2) **Retarders**. Chemicals that increase the setting time of a cement slurry.
3) **Extenders**. Chemical substances for reducing the density of cement slurry, or reducing consumption of dry cement to produce a unit volume of cement slurry. It should be noted that in this category there are additives which allow achieving both goals at the same time.
4) **Weighting agents**. Chemical substances for increasing the density of cement slurry.
5) **Cement slurry viscosity modifiers (dispersants).** Chemical substances enabling to regulate rheological properties of cement slurries, as a rule, the effect of this category of reagents is aimed at reducing viscosity.

Oil and Gas Well Cementing for Engineers, First Edition. Baghir A. Suleimanov, Elchin F. Veliyev, and Azizagha A. Aliyev.
© 2023 John Wiley & Sons Ltd. Published 2023 by John Wiley & Sons Ltd.

6) Additives to control fluid-loss of cement slurry
7) Additives to prevent loss of circulation during cementing
8) **Special additives**. This class of reagents includes a wide range of reagents performing various functions from foam suppressants to various fillers to increase the strength of cement. In fact, all chemical reagents that cannot be classified in one of the previous classes belong to this category.

Each of the following categories is discussed below in more detail.

3.2 Accelerators

Cement setting accelerators are used in cementing wells at shallow depths with low reservoir temperatures to reduce the setting time of cement slurry and ensure early strength development. Accelerators are also used for leveling the effect of additives causing excessive increase of cementing slurry setting time, such as viscosity modifiers and additives regulating fluid loss.

Accelerators affect the following processes:

- Reducing the setting time (I and II stages of cement hydration).
- Accelerating hardening of cement (III and IV stages of cement hydration).

Inorganic salts are often used as accelerators. The most common substances used as accelerator are chloride salts. Salts with the effect of accelerating the setting time of the cement mortar also include silicates, sulfates, hydroxides of potassium and ammonium.

Edwards and Angstadt [3] proposed the following classification of cations and anions according to their effectiveness as accelerators for Portland cement:

$$Ca^{2+} > Mg^{2+} > Li^+ > Na^+ > H_2O$$

$$OH^- > Cl^- > Br^- > NO_3^- > SO_4^{2-} = H_2O$$

When using salts as accelerators, salt concentration is critical (see Table 3.1). Thus, for example, calcium chloride ($CaCl_2$), is one of the most effective and affordable accelerator and is usually used at concentrations of 2–4% by weight of cement (BWOC), but when the concentration increases to 6% and higher it leads to premature shrinkage of the cement slurry.

In contrast, sodium chloride is very effective as a accelerator at concentrations up to 15% of the weight of the mixing water (BWOW), but has almost no effect at concentrations between 15% and 20%, and exceeding the concentration of 20% leads to an extension of the setting time (see Figure 3.1). In fact, at excessive concentrations sodium chloride (NaCl) performs the opposite role – retarding the

Table 3.1 Effect of calcium chloride on thickening time and compressive strength.

CaCl$_2$ (%BWOC)	Thickening time			Compressive strength (psi)								
	91°F/33°C (h:min)	103°F/39°C (h:min)	113°F/45°C (h:min)	60°F/16°C			80°F/27°C			100°F/38°C		
				6h	12h	24h	6h	12h	24h	6h	12h	24h
0	4:00	3:30	2:32	—	60	415	45	370	1260	370	840	1780
2	1:17	1:11	1:01	125	480	1510	410	1020	2510	1110	2370	3950
4	1:15	1:02	0:59	125	650	1570	545	1245	2890	1320	2560	4450

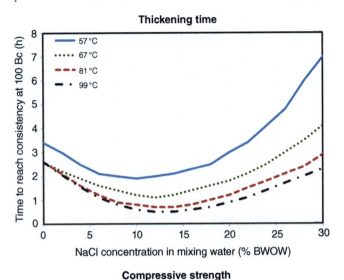
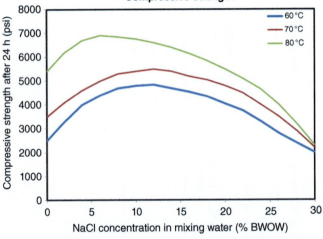

Figure 3.1 Influence of sodium chloride on thickening time and compressive strength.

setting time. In practice, the use of sodium chloride (NaCl) is recommended only when calcium chloride (CaCl$_2$) cannot be used.

Mineralization of mixing water has a strong influence on the setting time, so the use of seawater will inevitably increase the time of setting of cement slurry, as the NaCl content in seawater can be up to 2.5 wt%. In case of lack of fresh water source, the seawater application is the most widespread practice while carrying out cementing works at offshore oilfields. It should be noted, however, that application of seawater requires thorough laboratory tests of the cement slurry, as the complex chemical composition of seawater contains salts and chemical reagents of different nature that affect the cement slurry. In particular, the magnesium content is of great importance.

Significant heat release during the first hours of hydration when using $CaCl_2$ should also be noted, so the temperature of the casing, cement ring and adjacent zones of the formation may increase by 30 °C.

Accelerators that do not contain chlorine anions are classified as a separate group. For the first time the use of such accelerators is connected with civil construction. In particular with the construction of reinforced concrete structures since the use of accelerator containing chloride led to metal corrosion. During the construction of wells with the use of accelerators containing chloride, corrosion of the casing is also not uncommon.

Under normal conditions, the casing is protected against corrosion (passivated) due to the high pH value of the surrounding cement ring. A thin protective film of gamma iron oxide (γ-Fe_2O_3) is formed on the steel surface. The protective film prevents iron cations (Fe^{2+}) from entering the electrolyte and acts as a barrier to prevent oxygen anions (O^{2-}) from coming into contact with the steel surface of the casing. If passivation is compromised, casing corrosion occurs at a very high rate. This can be caused by factors such as a drop in ambient pH or by aggressive ions, such as chlorides, penetrating the casing/cement interface.

Another side effect of using accelerators with chlorine content is a significant release of heat, which, given the different coefficients of thermal expansion of the casing and cement, leads to the formation of the so-called micro-annulus. In fact, when hydration heat dissipates, casing shrinkage proceeds faster than cement ring shrinkage and a gap is formed between them. To date, this phenomenon has not been well studied and requires additional research.

The above-mentioned reasons were the basis for the development and application of accelerators that do not contain chlorine anions. Researchers considered many inorganic and organic compounds, including carbonates, silicates, aluminates, sulfates and hydroxides of alkali metals, as well as nitrates, nitrites, thiosulfates, formates as an alternative to calcium chloride. However, only a small fraction of these chemical compounds have been shown to be sufficiently effective accelerator when used in downhole applications.

Most accelerators that do not contain chlorine anions are complex compounds containing several chemical components. Let us consider the most common accelerators of this type.

Calcium formate ($Ca(HCOO)_2$), was patented as an accelerator in 1965. It is available in powder form, has poor solubility in water. The concentration of the additive is 1–2% by weight of dry cement. Calcium formate accelerates the process of hydration and as a consequence the setting of all classes of Portland cement. However, this reagent has a weak effect in the first 24 hours, which in practice is evened out by the addition of sodium nitrite.

Calcium nitrite (Ca(NO$_2$)$_2$), was patented as an accelerator in 1965. It has a good solubility in water, which is basically the main reason for using an aqueous solution of calcium nitrite rather than a powder form. Calcium nitrite is also an effective metal corrosion inhibitor, while making the strength of the cement stone comparable to the case of calcium chloride.

Calcium nitrate (Ca(NO$_3$)$_2$) combined with triethanolamine is an effective accelerator, but effectiveness varies widely depending on cement grade. Technical calcium nitrate is a mixture of calcium nitrate and ammonium hydrates and serves as both a setting accelerator and a corrosion inhibitor.

Triethanolamine (N(C$_2$H$_4$OH)$_3$), accelerates the reaction between C$_3$A and gypsum, and at concentrations of 0.1–0.5% by weight of dry cement, setting of the cement mortar can occur within minutes at ambient temperature. However, the setting of cement stone strength is much slower, as triethanolamine strongly slows down the hydration of C$_3$S and C$_2$S. For this reason, the compound has rarely been used as an accelerator in recent years.

3.3 Retarders

Cement slurry setting retarders are additives providing an increase in cement setting time in order to provide the time required for the execution of cementing jobs. There is still no consensus on the mechanism of action of these additives. The main difficulty in this issue is that it is necessary to take into account both chemical properties of the additive itself and of the cement phase (silicate or aluminate), which it affects. Nevertheless, researchers distinguish four theories below to explain this mechanism:

1) **Adsorption theory**. According to this theory, the main active force is the adsorption of setting retardants on the surface of the hydration products, which creates a kind of barrier with water and thus slows down the further hydration process.
2) **Precipitation theory**. According to this theory, cement setting retarder reacts with calcium ions, hydroxyl ions or both simultaneously to form an insoluble and impermeable layer around the cement grains, preventing contact with water and consequently retarding the hydration reaction.
3) **The theory of nucleation**. According to this theory, the cement setting retarder is adsorbed on the nuclei of hydration products, blocking their growth.
4) **The theory of complexation**. According to this theory, the cement setting retarder chelates calcium ions, preventing the formation of hydration product nuclei.

Apparently, all of the above retardation mechanisms to some extent occur. Despite the lack of consensus on the mechanism of action of cement setting retarders, the main chemicals that provide an increase in cement setting time are classified clearly enough.

3.3.1 Lignosulfonates

The most commonly used setting retarders are sodium and calcium salts of lignosulfonic acids. Lignosulfonates are a byproduct of pulp and paper production, which leads to an increased content of impurities in the form of various saccharide compounds. Lignosulfonates purified from impurities have a much lower retarding capacity, which is why a number of researchers are inclined to explain the retarding effect of lignosulfonates by the presence of impurities. The chemical composition of these impurities also varies widely, including such low-carbohydrate compounds as xylose, arabinose, hexoses (mannose, glucose, fructose, rhamnose and galactose), xylonic and gluconic acids. Lignosulfonic setting retarders are effective with all classes of Portland cement, the average concentration in the cement mortar is from 0.1% to 1.5% by weight of dry cement. Depending on the carbohydrate content of the lignosulfonate retarders and cement grade, they can be used quite effectively at Bottomhole Circulating Temperature (BHCT) up to 122 °C (see Figure 3.2). When the effective temperature range of an application needs to be increased, sodium borate is usually added additionally, thus increasing the upper temperature limit of the application to 315 °C (BHCT).

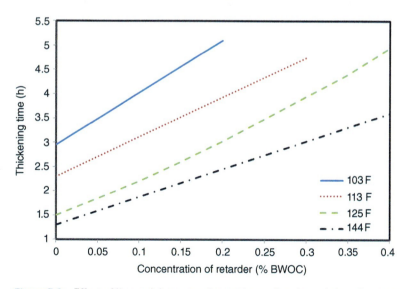

Figure 3.2 Effect of lignosulphate retarder on the setting time of class G cement.

The mechanism of action of this class of retarders is explained by a combination of adsorption and nucleation theories. Lignosulfonate setting retarders affect the hydration kinetics of C_3S and C_3A, with a more significant effect on C_3S. It should be noted that this type of retarders is particularly effective when used with cements with low C_3A content.

3.3.2 Hydroxycarboxylic Acid

Gluconate, glucoheptonate salts and tartaric acid are the most widely used chemical compounds in this category, having a powerful retarding effect at bottomhole circulation temperature (BHCT) below 93 °C they significantly increase the setting time of cement slurry, which in turn can cause a number of related complications. However, this type of setting retarder has been shown to be quite effective at higher bottomhole circulation temperatures (BHCT) up to 150 °C.

Citric acid is another hydroxycarboxylic acid that has a severe retarding effect. It is also an effective cement dispersant, with average concentrations in cement mortar ranging from 0.1% to 0.3% by weight of dry cement. The retarding effect of hydroxycarboxylic acids and their salts is usually due to the presence of α- or β-hydroxycarbon groups ($HO-C-CO_2H$ and $HO-CC-CO_2H$, respectively), which can strongly chelate metal cations. Like lignosulfonates, hydroxycarboxylic acids are more effective when used with cements with low C_3A content.

3.3.3 Saccharide Compounds

Saccharide compounds are very effective setting retarders of cement mortars. The mechanism of action of this group of reagents depends on the degree of decomposition by alkaline hydrolysis, due to which saccharide compounds are converted into saccharine acids containing α-hydroxycarbonyl groups ($HO-C-C=O$), which are strongly adsorbed on the surface of the C–S–H phases.

3.3.4 Cellulose Derivatives

These compounds are polymers and are mainly polysaccharides derived from wood or other plants. They are stable in the alkaline environment of cement solutions. The retardation of setting is due to the adsorption of the polymer on the surface of the hydrated cement. The active centers of adsorption are ethylene oxide links and carboxyl groups.

The most common setting retarder of this class of reagents is carboxymethyl-hydroxyethylcellulose (CMHEC). The upper temperature limit of application usually does not exceed 121 °C. It should be noted, however, that there are a number of side effects when used, such as an increase in the viscosity of the mortar and this reagent is used more as an additive to control water loss.

3.3.5 Organophosphonates

Phosphomethylated compounds containing quaternary ammonium groups as well as *N*-phosphonomethyliminodiacetic acid are effective retarders of cement setting at circulating temperatures up to 232 °C. It should be noted that at temperatures above 110 °C methylene phosphonic acid derivatives are not effective and the upper temperature threshold of application is increased by adding borate salt. With the appropriate phosphonate/borate ratio the gelling time increases considerably, the set strength of the cement stone is accelerated and the upper temperature threshold of application reaches 232 °C. Low sensitivity to slight changes in cement composition and lower viscosity of cement mortars of high density are also significant advantages of using these compounds as retarders of cement setting. The mechanism of action of organophosphonates is explained by the adsorption theory.

3.3.6 Inorganic Compounds

Many inorganic compounds also retard the hydration of Portland cement. The main classes of these materials are listed below:

- **Acids and salts**. boric, phosphoric, hydrofluoric and chromic acids.
- Sodium chloride at a concentration of more than 20% by weight of the mixing water.
- Zinc and lead oxides.

Sodium tetraborate decahydrate (borax: $Na_2B_4O_7 - 10H_2O$) should also be noted, which allows to significantly increase the upper threshold of most lignosulfonate stiffeners up to 315 °C. However, this additive negatively affects the effectiveness of cellulose and polyamine additives that reduce water retention.

3.4 Extenders

This class of additives aims to achieve the following two objectives:

- **Reduction of cement slurry density**. Cement slurries without any additives, regardless of cement class used, have rather high density, which leads to significant increase of hydrostatic pressure at large cementing intervals and may lead to loss of cementing slurry circulation during cementing of weakly cemented formations due to formation fracturing. The possibility of reducing the number of cementing intervals is not unimportant either.
- **Increase of cement slurry volume**. Additives increase the volume of cement slurry, which allows to reduce the consumption of Portland cement to obtain the required volume of cement slurry.

The extenders are divided into three different groups depending on the mechanism of action:

- **Extenders that increase the water content of the cement mortar.** The most common compound in this group is clay. Clay absorbs water very well, increasing in size many times, and the increased water content naturally leads to a decrease in the density of the cement mortar. Since the additional amount of water introduced into the cement mortar is actually bound to the clay particles, the free water content remains low and the cement mortar is homogeneous and stable.
- **Extenders with low density.** The mechanism of action of this group of reagents is based on adding compounds with lower density to cement mortar compared to Portland cement ($3200\,kg/m^3$). In fact, the main difference from the first group is that the first group for this purpose uses water ($1000\,kg/m^3$), while the second group uses different, as a rule, bulk materials.
- **Gas-based extenders.** This group of compounds includes various gases, the most common of which are nitrogen and air. The mechanism of action is almost identical to that described above, but in this case the filler is not water or bulk material, but gas. The essential difference is that the injection of gas into the cement mortar creates foam-cement, which has a number of specific properties.

In practice, when formulating cement mortar compositions, only one group of relieving additives is rarely used, more often they are combined. Table 3.2 shows the most common facilitation additives.

3.4.1 Clays

Clay minerals are aqueous silicates, aluminosilicates and ferrosilicates that have a layered and pseudolayered crystalline structure. The mineral particle is formed from layers consisting of silica-oxygen tetrahedrons and alumohydroxyl octahedrons. Layers form a set called packages; depending on the set of packages in each layer, clay minerals are divided into different groups. For example: kaolinite group, hydromica group, smectite group, etc.

The most used clay mineral is bentonite. Bentonite consists of 85% smectite ($NaAl_2(AlSi_3O_{10})(OH)_2$) and swells up to 16 times when hydrated. When it swells, it forms a dense gel that prevents further moisture penetration. That is why bentonite is often called gel. There are two types of bentonite:

- calcium, with a low degree of swelling;
- sodium, with a high degree of swelling.

Table 3.2 Extenders.

Extender	Density reduction range of cement slurry (ppg) 6–16	Advantages
Bentonite	11.5–15	Helps control the fluid loss of the cement slurry
Fly ash	13.1–14.1	Increases resistance to corrosion
Sodium silicate	11.1–14.5	Effective at low concentrations and when using seawater for cement mixing
Microspheres	8.5–15	Increases cement stone strength and thermal stability. Reduces the permeability of the cement stone
Foam cement	6–15	Increases the strength of the cement stone by reducing its permeability

Bentonite concentration in cement slurry, as a rule, does not exceed 20% by weight of dry cement, since even at a concentration exceeding 6% the viscosity of cement slurry increases considerably, which requires adding another group of cement additives, viscosity reducers. According to API/ISO recommendations it is required to add 5.3% extra amount of water by weight of dry cement for every 1% of bentonite regardless of cement class. It should also be noted that increasing viscosity is not the only undesirable effect of adding bentonite to cement slurry, since increasing the concentration of bentonite increases the permeability of cement stone and as a consequence, reduces the sulfate resistance. Nevertheless, in view of the fact that bentonite actually binds water during hydration, water release of cement slurry significantly decreases and this property is often used by engineers while preparing chemical compositions of cement slurry. Table 3.3 shows the dynamics of such parameters as mixing water volume, density and solution output depending on the bentonite content.

Due to the reduction of the degree of bentonite swelling with increasing salinity of the mixing water, in practice, it is pre-hydrated in fresh water and an aqueous bentonite solution is added to the cement slurry. For complete hydration of bentonite 30 minutes is sufficient, which in fact does not affect the duration of preparation of cement slurry. In addition, the efficiency of hydrated bentonite compared to dry powder is 4 to 1, which means that only a 5% concentration of hydrated bentonite is required in order to achieve identical solution parameters corresponding to a 20% non-hydrated bentonite content. As noted above, the hydration of bentonite is possible only in low saline water, and if highly mineralized water is used, bentonite is usually replaced by another clay mineral attapulgite $((Mg,Al)_5Si_8O_{22}(OH)_4 - 4H_2O)$. In contrast to bentonite, this mineral does not improve the water loss of cement slurry and is banned in some countries.

Table 3.3 Influence of bentonite on cement slurry properties.

Concentration (%)	Mix water volume (gal/sack)	Density (ppg)	Cement slurry yield (ft^3/sack)
0	4.97	15.8	1.14
2	6.17	15.0	1.31
4	7.36	14.4	1.48
6	8.56	13.9	1.65
8	9.76	13.5	1.82
10	10.95	13.1	1.99
12	12.15	12.7	2.16
16	14.55	12.3	2.51
20	16.94	11.9	2.85

3.4.2 Sodium Silicate

The mechanism of action of this group of relieving additives is based on the formation of calcium silicate gel as a result of the reaction of silicates with lime or calcium chloride in the cement. The resulting gel absorbs additional water, increasing its content in the mortar. The water is in a bound state, which prevents excessive water loss due to the excessive amount of water. It should be noted that addition of sodium silicate leads to increase of viscosity of mortar and acceleration of setting, reducing the efficiency of additives regulating water release and setting retarders.

Sodium silicates are available in liquid and solid (bulk) form.

Solid sodium silicate, (Na_2SiO_3-sodium metasilicate), is usually mixed directly with cement powder. Gel formation does not occur in fresh water; calcium chloride is added to the mixing water in advance for this purpose. The recommended concentration range for sodium metasilicate is 0.2–3.0% by weight of dry cement. These concentrations ensure reduction of cement slurry density to 1740 and 1320 kg/m³, respectively.

Liquid sodium silicate (Na_2O – (3–5) SiO_2 – liquid glass) is mixed with mixing water. If fresh mixing water is used, calcium chloride is added beforehand. In seawater, the gel is formed without calcium chloride by the interaction of sodium silicate with divalent cations. The recommended concentration range of liquid sodium silicate is 0.2–0.6 gal/sack. These concentrations provide lower density of cementing solution up to 1700 and 1380 kg/m³, respectively.

3.4.3 Pozzolans

Pozzolans are rocks consisting of loose products of volcanic eruptions, and are chemically siliceous or siliceous/alumina-containing materials.

There are two types of pozzolans:

1) natural pozzolans consisting of volcanic ash and diatomaceous earth (kieselgur)
2) artificial pozzolans such as fly ash.

Pozzolans in a finely dispersed form with sufficient humidity react with calcium hydroxide to form compounds with cementitious properties. This property of pozzolans allows them not only to reduce the density of cement slurry, but also to increase the strength of the cement stone, reacting with the calcium hydroxide released during the hydration of cement. Thus, the hydration of 94 pounds (weight of a standard bag of Portland cement) releases up to 23 pounds of calcium hydroxide ($Ca(OH)_2$), given that calcium hydroxide is well soluble in water eventually it may be dissolved and removed by water contact with cement, which naturally leads to a decrease in the strength of the cement stone. However, if pozzolan is present in the cement composition, the silica in the cement composition reacts with the released calcium hydroxide to form the so-called "secondary" calcium hydrosilicate (C–S–H). Calcium hydrosilicate increases strength and reduces the permeability of the cement stone (0.001 mD). Low permeability of cement stone prevents penetration of aggressive formation fluids. Below we will consider some of the most common in the oil industry compounds from this group.

3.4.3.1 Diatomaceous Earth (Kieselgur)

Diatomaceous earth consists of the remains of diatom algae. The main component of diatomaceous earth is opal, an amorphous form of silica containing up to 10% water. For use as a facilitating additive, diatomaceous earth is ground to the size of Portland cement, which increases the specific surface area and correspondingly the amount of mixing water required. The diatomaceous earth has the same effect as the addition of bentonite, giving similar properties to the cement mortar, while increasing its viscosity to a much lesser extent. A distinctive feature compared to the use of bentonite is also a higher strength of the cement stone. The main disadvantage of diatomite is its cost, which is many times higher than the cost of bentonite. The recommended concentration range is up to 40% by weight of dry cement, which reduces the density of grouting mortar to 1320 kg/m^3.

3.4.3.2 Fly Ash

Fly ash is a product of coal combustion (for example: in thermal power plants), formed as a result of solidification of molten particles in the flue gases, it has almost spherical shape. This product due to its high dispersibility has a specific surface area comparable to that of Portland cement. The main components are: silicon dioxide, aluminum and iron oxides, lime, etc. However, it should be noted that the composition of the fly ash can vary quite widely depending on the composition of the coal burnt.

According to ASTM classification depending on the chemical composition the following three classes of fly ash are distinguished: N, F, and C (see Table 3.4).

Class N and F fly ashes are formed by burning anthracite (bituminous coal). Fly ash of class C, formed when burning lignite (lignite), contains less silicon, and in some cases more than 10% of lime. For cementing wells class F fly ash is most often used, as a rule, up to 2% of bentonite is also added to this composition to prevent excessive content of unbound water. Application of class C ash dust, due to its high lime content and great variability of chemical composition, significantly complicates the process of composing the cement slurry and results in the necessity to perform multiple laboratory tests, which hinders its wide practical application. Nevertheless, some fly ash of this class has good cementing properties and can be quite well used as the main plugging material in shallow wells with bottom hole circulation temperature up to 50 °C.

Table 3.4 Chemical composition requirements for fly ash according to ASTM classification.

	Fly ash class		
	N	F	C
Min. content: silicon dioxide (SiO_2) + aluminum oxide (Al_2O_3) + iron oxide (Fe_2O_3) (%)	70	70	50
Max. sulfur oxide VI (SO_3) content (%)	4	5	5
Max. moisture content (%)	3	3	3
Max. combustion loss (%)	10	12	6

3.4.3.3 Lightweight Cementing Slurries

In the market there is quite a wide choice of lightweight cements, consisting of fine Portland cement and different siliceous compounds, which are complete plugging materials with the density of 1430–1640 kg/m³. Although this group of compounds is not actually an additive to cement (because of the chemical composition), most researchers tend to classify and describe it when describing pozzolans. Authors, also adhere to a similar classification.

3.4.3.4 Silica (Silicon Dioxide, Quartz)

The following three forms of silica are used in well cementing operations:

- α-quartz
- microsilica (amorphous condensed microsilica)
- colloidal dispersions of silica

α-quartz – used to prevent deterioration of the strength of the cement stone in thermal wells. According to the size of the particles we can distinguish the two most common products of this group:

- quartz sand with an average particle size of 100 μm
- quartz dust with an average particle size of 15 μm.

Of course α-quartz of other sizes exists and is used in practice, e.g. in special cements with adjustable particle size distribution, but due to their high cost, the use of such materials as extending agents is not feasible.

Microsilica (amorphous condensed microsilica) is a byproduct of silicon production and silicon alloys. Individual particles are glassy amorphous microspheres. The average particle size is typically between 0.1 and 0.2 μm, which is about 50–100 times smaller than the particles in Portland cement or fly ash. The specific surface area ranges from 15 000 to 25 000 m²/kg. The high specific surface area increases the amount of mixing water required and practically reduces to zero water separation in the mortar. Due to this property, it also significantly reduces the water loss of cement mortar, permeability of filtration crust and cement stone. Perhaps the most significant disadvantage of using this additive is only the need to add high concentrations of cement setting retarders. The standard concentration of microsilica in cement slurry is 15% by weight of dry cement, however, it should be noted that in rare cases it is possible to increase this value to 28%.

Colloidal dispersions of silica, are aqueous sols of pure amorphous silicon dioxide and extremely small amounts of sodium hydroxide. Like microsilica, colloidal silica particles are spherical in shape, but the particle size is about an order of magnitude smaller (0.05 μm), resulting in a high specific surface area of 500 000 m²/kg.

3.4.4 Lightweight Particles

Particle-based lightweighting additives reduce the density of the cement mortar due to the lower density of the filler particles compared to the Portland cement particles. As a rule, such additives are inert in relation to Portland cement. The most common materials in this category are:

- Expanded perlite
- Gilsonite
- Powdered carbon
- Microspheres (ceramic, glass)

3.4.4.1 Expanded Perlite

Perlite is crushed volcanic glass which expands when heated to the melting point. Density of expanded perlite is 124–155 kg/m³ which allows to lower density of cement slurry to 1440 kg/m³. In practice, to prevent segregation of perlite particles, a small amount of bentonite from 2% to 4% by weight of dry cement is added to the mortar. Having quite a porous structure perlite is unstable to pressure increase and its application, as a rule, is limited to not deep wells. Today perlite is rarely used in practice due to the high variability in density depending on downhole pressure.

3.4.4.2 Gilsonite (Asphaltum)

Gilsonite is a natural asphaltite resinous hydrocarbon, often referred to as natural asphalt. It does not require a large amount of mixing water (7.5 l per 0.02 m³) and forms a cement stone with a relatively high compressive strength. Concentration in cement mortars can reach 50% by weight of dry cement, the density of cement mortar, while reducing to 1440 kg/m³. However, at such high concentrations the pumpability of the cement slurry is considerably impaired. Just as with the use of perlite, a small amount of bentonite is added to the mortar to maintain homogeneity. It is worth to note, that it is not recommended to use gilsonites at downhole static temperatures above 149 °C, as already at temperatures above 116 °C some softening of gilsonite particles is observed. Gilsonite is also often used to prevent loss of circulation.

3.4.4.3 Powdered Carbon

According to its characteristics, powdered carbon is almost identical to gilsonite, has a coarse-grained composition (more than 50% of particles have a size in the range from 1.68 to 3.36 mm) and is used for prevention of circulation loss. The essential difference is high melting point (538 °C), which allows its application in high-temperature wells. Concentration in cement slurry is in the range from 10% to 26% by weight of dry cement, allowing to decrease density of cement slurry up to 1430 kg/m³. In order to maintain mortar uniformity bentonite is also added to the composition.

3.4.4.4 Microspheres

Microspheres are small gas-filled spheres with a density of 120–900 kg/m³, which allow reducing the density of the cement slurry to 1020 kg/m³. According to their composition microspheres are divided into two types: glass and ceramic. Microspheres are unstable to pressure rise, which limits their application in deep wells. An important factor is a uniform distribution of microspheres over the volume of the cement mortar, so microspheres are usually added to dry cement, not to the mixing water

Glass microspheres are produced from borosilicate glass, density varies from 120 to 800 kg/m³, average particle size is 30–40 μm. Classification into different grades is carried out according to the wall thickness of microspheres. With increasing wall thickness naturally increases and the density, but also increases the value of the maximum allowable pressure of application. So the most dense grades can withstand pressure up to 70 MPa, although on average this value for most grades with standard wall dimensions is 35 MPa. Glass microspheres are much more expensive than their ceramic counterparts, so their use is not widespread.

Ceramic microspheres are obtained from ash produced by coal-fired power plants. Their composition varies, but the main components are silicon dioxide and aluminum oxide, the average particle size varies in a wide range from 20 to 500 μm. The thickness of the shell is about 10% of the particle radius. The gas composition inside the microspheres is a mixture of CO_2 and N_2. Ceramic microspheres have a higher density compared to their glass counterparts of 600–900 kg/m³ with a bulk density of 400 kg/m^3, which leads to the need to increase their concentration in the cement mortar to obtain the target density.

3.4.5 Gas Based Extenders

3.4.5.1 Nitrogen

When this technology is used, gas is injected directly into the cement slurry, which leads to the formation of foamed cement and, consequently, to a reduction in density. Nitrogen injection makes it possible to achieve a cement slurry density of 0.84 kg/m^3. It should be noted that a number of requirements are imposed on the composition of the cement slurry to achieve high strength and low permeability of the cement stone.

3.5 Weighting Agents

To prevent drilling complications such as fluid and gas blowouts and wellbore collapse while drilling through high-pressure formations, it is critical to maintain a high hydrostatic pressure in the drilling fluid. In some such wells, drilling fluid densities often exceed 2000 kg/m^3. When cement slurry displaces drilling mud, the former actually replaces the latter in the wellbore, and the density of the cement slurry must be comparable to that of the drilling fluid to ensure proper hydrostatic pressure. It should be reminded that the density of cement slurries does not exceed 1965 kg/m^3 (H-grade cement). The simplest solution is to reduce water content in the slurry, but it inevitably leads to worsening of pumping ability of the slurry and necessity of using viscosity modifiers. But this method also allows increasing mortar density only up to 2160 kg/m^3, requiring careful selection of mortar constituents to provide stable rheological indicators, cement stone density and acceptable water yield values, which in practice is a very laborious and complicated task. More practical and technologically easy solution is to add fillers with high density. This group of additives must meet the following requirements:

- the average particle size should be comparable to Portland cement particles, because if the particle size of the weighting additive is larger, it will lead to their precipitation in the suspension, and smaller ones will lead to a significant increase in the viscosity of the mortar,
- the amount of mixing water needed should be relatively low,
- the weighting agent must be chemically inert and compatible with other additives
- do not cause a significant decrease in the strength of the cement stone due to a decrease in the cement content per unit volume

The most common weighting additives for cement mortars are ilmenite, hematite, hausmannite and barite (see Table 3.5).

3.5 Weighting Agents

Table 3.5 Physical properties of cement slurry weighting agents.

Additive	Specific gravity	Absolute volume (gal/ft)	Color	Additional mixing water requirement (gal/ft)
Ilmenite	4.45	0.027	Black	0.00
Hematite	4.95	0.024	Red	0.0023
Barite	4.33	0.028	White	0.024
Hausmannite	4.84	0.025	Reddish brown	0.0011

3.5.1 Ilmenite (Iron Titanium Oxide)

Ilmenite ($FeTiO_3$ or $FeTiO_2$ composition is not constant), granulated black material with a density of 4500–5000 kg/m^3, weakly magnetic, is a source of obtaining valuable titanium alloys. The use of ilmenite as a weighting additive allows to increase cement mortar density up to 2400 kg/m^3, but because of coarse-grained granulometric composition there is a high risk of precipitation. Strict control of the viscosity of the cement slurry is necessary to prevent such a development.

3.5.2 Hematite

Hematite (Fe_2O_3) iron ore, is the most widely used weighting agent, the density of the mineral ranging from 4900 to 5300 kg/m^3. Despite the high density, the small average particle size prevents settling in the cement mortar and promotes being in suspension in the mortar. However, the fine dispersion has a negative effect on the viscosity of the cement slurry and leads to its increase, which causes the need to use cement slurry viscosity modifiers (i.e. dispersants). The table shows the dynamics of cement slurry density change depending on hematite concentration (see Table 3.6). In practice, it is also possible to obtain higher density values, but this significantly complicates the chemical composition of cement slurry.

3.5.3 Hausmannite

Hausmannite ($Mn_2 + Mn_{23} + O_4$) is a reddish-brown mineral with a density of 4700–4800 kg/m^3. Being finer (average particle size is 5 μm) in comparison to other weighting additives, hausmannite does not precipitate in the cement mortar and can be added directly to the mixing water. However, adding hausmannite in case of excessive dosage may result in a significant increase in viscosity and reduction of cement slurry gelling time. Combined use of the additive with hematite in a cement slurry composition makes it possible to increase cement slurry density

Table 3.6 Hematite slurries with H-grade cement.

Hematite (ft/sack)	Water (gal/sack)	Weight (ppg)	Slurry yield (ft³/sack)
0	4.29	16.47	1.05
10	4.50	17.00	1.11
20	4.60	17.68	1.16
30	4.66	18.15	1.20
40	4.75	18.70	1.24
50	4.81	19.15	1.28
60	4.90	19.60	1.33

to 2640 kg/m³. It is available both in powder and liquid form, the latter one being widely used at offshore fields.

3.5.4 Barite

Barite ($BaSO_4$) a white powder with a density of 4230 kg/m³ has been widely used as a weighting agent in drilling mud for many decades. However, application in cement slurries is not as effective as other weighting additives. Barite, due to its coarse particle size distribution needs additional amount of mixing water (about 22% of barite weight), which finally increases water content of cement slurry and as a consequence cement stone density and strength values drop. The problem is partially solved by adding a dispersant to the composition. Barite can increase the density of cement mortar up to 2280 kg/m³.

3.6 Dispersants

Cement slurries are highly concentrated suspensions, i.e. coarse disperse system with solid disperse phase (cement) and liquid dispersion medium (water). It should be noted, that the solid phase is understood as the whole set of solid additives and cement, included in slurry, and the liquid dispersion medium is mixing water and all liquid additives. Rheological properties of such system depend on the following factors:

- Rheological properties of the liquid dispersion medium.
- The volume fraction of the solid phase.
- Interaction of particles of the dispersed phase.

Aqueous phase of cement mortar. Rheological indexes of this phase vary quite widely, as they depend both on chemical composition of additives included in cement composition and on temperature.

Volume fraction of the solid phase is the ratio of the total volume of particles of the solid phase to the volume of cement mortar, this index ranges from 0.2 to 0.7 for lighter and heavier cements, respectively.

Particle interaction primarily depends on the surface distribution of charges and steric effects caused by organic molecules adsorbed on the surfaces of the solids.

Without viscosity modification, most cement mortars would not have the necessary rheological properties. Additives used to adjust the viscosity of cement slurry in the oil industry are called dispersants, but it is not uncommon to use the terminology of the construction industry, where cement slurry viscosity modifiers are divided into two main categories: plasticizers and superplasticizers. It is this classification that is used to divide dispersants by chemical composition:

- **Plasticizers** – include the following chemical compounds: lignosulfonates, citric, tartaric and salicylic acids, but having a strong effect of retarding the setting of cement, plasticizers have gained more use in this context.
- **Superplasticizers** – include predominantly sulfated polymers such as polymelamine sulfonate, polynaphthalene sulfonate, etc.

Lignosulfonates as dispersants are more common in drilling fluid formulations than cement-based formulations. Lignosulfonates are also effective cement setting retarders and their use at low circulating temperatures downhole is not desirable. At high temperatures, the rate of cement hydration increases, which somewhat mitigates the retarding effect of lignosulfonates. This group of compounds is extremely sensitive to cement composition and strong fluctuations in rheological properties are possible depending on cement class.

Polynaphthalene sulfonate, due to its affordability is the most common and widely used dispersant for cement slurries. However, due to its toxicity and adverse environmental effects (bioaccumulation) on a number of marine algae, the use of this compound is legally prohibited in some marine ecosystems. Polynaphthalene sulfonate is available both in powder form and as a 40% water solution. The standard concentration in cement mortar ranges from 0.2% to 4% by weight of dry cement. As the mineralization of mixing water increases, so does the concentration of polynaphthalenesulfonate. The upper temperature threshold for the use of this compound is 204 °C.

Polymelamine sulfonate is a fairly effective dispersant, however, due to its high cost compared to polynaphthalene sulfonate, is more widely used in the construction industry compared to the oil industry. Available in powder form and 20% or 40% aqueous solution. As a rule, the concentration in cement slurry is

0.4% by weight of dry cement, but these values can vary greatly depending on the grade of cement. The upper temperature threshold for use of this compound is 121 °C.

In the dispersant application there are two main and most critical factors from the point of view of isolation efficiency:

- **Water separation**, that is, the release of unbound so-called *"free water."* The gel formed during cement hydration and the walls of the narrow borehole annulus should theoretically act as a kind of a carrier agent for non-hydrated cement particles. However, such situation is rare in practice, the above-described phenomena have no special influence on the particles suspended in the cement slurry and the weight of cement particles is transferred to the layers below, squeezing water out of them. The released water rushes upwards through the annular space, forming a layer of unbound or so-called "free water." It should be noted that the permissible level of "free water" in a solution is ruled by regulatory documents.
- **Sedimentation**, that is heterogeneity of cement slurry, caused by different rates of sedimentation of large and small cement particles. Researchers have no consensus about the mechanism of this phenomenon, but there are three main hypotheses explaining the mechanism of cement slurry sedimentation:
  – Uneven sedimentation rate of fine and coarse cement particles
  – Brownian motion. According to this hypothesis, Brownian motion prevents the deposition of small cement particles while not significantly affecting the large particles.
  – The resulting gel does not have enough carrying capacity to keep the large cement particles suspended.

Whatever the mechanism of this phenomenon, its essence remains the same: the cement slurry column has an uneven density and, consequently, the strength of the cement stone along the cementation interval will be different during hardening (see Figure 3.3).

It is not uncommon for both the deposition of cement particles and the release of excessive amounts of free water to occur simultaneously. To prevent these phenomena, it is critical to strictly control the concentration of dispersant in the solution, but since we are often talking about tenths of a percent in the field, this is difficult to achieve. For this purpose, antisegregation (stabilizing) additives are used. These allow to increase the flowability of the fluid to a level compatible with pumping conditions, i.e. leaving the fluid flowable enough to be pumped into the well. Such additives include bentonite, various water-soluble polymers, cellulose derivatives (hydroxyethyl cellulose, Welan gum), seawater, silicates, metal salts

3.7 Fluid Loss Agents

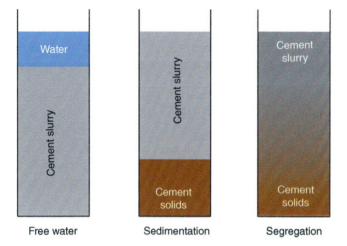

Figure 3.3 Three different processes for the sedimentation of cement slurry.

($NiCl_2$, $MgCl_2$). The effectiveness of stabilizing additives is evaluated in the laboratory as follows:

- the cement mortar to be tested is placed in a cylindrical mold for solidification
- after curing of the mortar, the cement mold is extracted and cut into equal parts
- the density of the samples is measured

The difference in density of the samples and is an indicator of the effectiveness of the stabilizing additive.

3.7 Fluid Loss Agents

Cement slurry fluid loss is the separation and filtration of the aqueous phase of the cement slurry into the formation under the action of differential pressure between the slurry and the formation. This process, in the absence of strict control, leads to a number of serious complications. As the water content in the solution decreases, both physical and chemical properties of the slurry and such indicators as viscosity, thickening time, strength, etc. change significantly. It is difficult to completely avoid fluid loss of cement slurry in practice, but a certain safe value for work is still allowed (less than 50 ml/30 minutes according to API standards). Efficiency of cement slurry prepared according to API recommendations without any additives usually exceeds 1500 ml/30 minutes, i.e. actually

30 times more than allowable values. To reduce fluid loss of cement slurries to permissible values, additives regulating fluid loss are used.

It should be noted that the exact mechanisms of action of this class of additives is unknown and they are usually explained by simultaneous course of several interrelated processes.

Under the action of differential pressure, the liquid phase rushes into the reservoir, leaving behind solid particles that precipitate on the surface of the reservoir and form a "filter cake." The mechanism of action of water shut-off additives is aimed either at reducing the permeability of the given crust or at increasing the viscosity of the liquid phase, or both processes simultaneously. There are two major classes of water-reducing additives: fine-dispersed solids and water-soluble polymers.

3.7.1 Particulate Materials

The first and most widespread product of this class, used to reduce fluid loss of cement slurry, was bentonite. The finely dispersed granulometric composition allows it to successfully penetrate into the filtration crust, clogging the pore channels and, as a consequence, reducing the permeability. In addition to bentonite such compounds as silica microdioxide, asphaltenes, thermoplastic resins, etc. are often used. Latex cement systems also have good fluid loss prevention properties. In recent years cross-linked polymer systems such as cross-linked polyvinyl alcohol microgels have also proven good. The additive is suitable for use in a wide temperature range, as it is almost chemically inert to cement accelerators. Additives of this class differ from additives based on water-soluble polymers in that the values of fluid loss of cement do not depend on time. There is a one-time release of filtrate into formation with formation of filtration crust, further loss of liquid is insignificant and for practical calculations can not be taken into account.

3.7.2 Water Soluble Polymers

Water-soluble polymers have been used as additives to reduce fluid loss in drilling fluids since the middle of the last century. Today, they are successfully used for this purpose in cement slurries. The main mechanism of action of this type of additives comes down to reduction of permeability of filtration crust and increase of viscosity of water phase of cement slurry. Moreover, the first mechanism of action is more effective, since at high polymer concentrations in cement slurry its viscosity increases significantly. For example, in order to decrease permeability of filter cake by 1000 times the viscosity of liquid phase should be increased by no more than 5 times, while replacing a low-molecular polymer with a high-molecular one the viscosity grows by more than 100 times. Water-soluble polymers form weakly bound colloidal aggregates in the cement mortar, which are stable enough to plug the pore channels of the filter cake. Such polymers may also adsorb to the surface of cement grains, reducing pore size. In all likelihood, both mechanisms are actually involved. In contrast to water-reducing additives in the form of solids, water-soluble polymers do not contribute to the formation of a thin and impermeable cement filtration crust. Instead, they simply reduce the rate of thickening of the filtration crust, and the process proceeds until the suspension dewaters and a thick filtration crust is formed.

3.8   Lost Circulation Prevention Agents

The main causes of cement slurry circulation failure include high cavernousness of the wellbore, presence of weakly cemented rocks and relatively narrow range of allowable pressure gradients (i.e. allowable cement slurry density is limited in a very narrow range). Cement slurry circulation disruption is usually preceded by an identical complication occurring during drilling of a given horizon. In fact, it rarely comes as a surprise to the cementing crew. To prevent such complications, a class of additives designed to prevent loss of circulation during cementing is used. The mechanism of these additives is to plug fractures and poorly cemented rock, resulting in increased formation strength and higher fracture pressure. The nature of this class of additives is generally inert to the Portland cement hydration process. Most of the compounds used for this purpose are various granular materials such as: gilsonite, granular coal, and ground nutshells. However, application of such materials as coarse-grained bentonite and corn cobs is also possible. In recent years technologies based on injection of cellophane flakes and addition of different fibrous materials into cement have become widespread. If the above-mentioned plugging agents are not effective, as a rule, thixotropic cements are used, which, getting into the absorption zones, form a gel, significantly reducing the permeability, due to the decrease of shear rate.

3.9   Special Cement Additives

A number of materials are added to cement slurries that do not fit into any general category. These include antifoaming agents (defoamers), cement rock strength additives, drilling fluid cleaners, and radioactive tracers.

3.9.1   Antifoaming Agents (Defoamers)

Many cement additives foam the cement slurry during mixing, which leads to a number of undesirable complications, such as premature gelation and changes in hydrostatic pressure in the well, since in the bottomhole conditions the increased pressure will lead to air compression and, consequently, increase the density of cement slurry. In order to prevent such complications anti foaming agents are used. Defoamers are divided into two classes according to the mechanism of action:

- **Compounds aimed at preventing foaming**. Defoamers in this class are added either to the cement or to the mixing water before the mortar is mixed. Adding anti foaming agents of this category after mixing mortar is not effective. The mechanism of action is based on chemical interaction with the foaming substance forming slightly soluble or insoluble compounds.
- **Compounds aimed at destroying the formed foam**. The mechanism of action of these additives is based on various physical processes, while they are chemically inert with respect to the foaming substances.

The most common defoamer due to its economic availability is polyethylene glycol. The effective concentration is usually less than 1% by weight of the mixing water. It belongs to the first class of antifoaming agents and is added before mixing the cement mortar. When added after mixing, on the contrary, it increases the stability of the foam.

The other most common blowing agent is a suspension of finely dispersed silica particles dispersed in various silicones. These products destroy the foam equally well when added before mixing the cement slurry as well as afterwards.

3.9.2 Strengthening Agents

This class of additives to cement is represented by various fillers that significantly improve the mechanical properties of the cement concrete. Today the range of compounds used for this purpose is quite extensive, from nanoparticles to various fiber materials. The traditional compounds in this group include various fiber products added to cement in concentrations ranging from 0.15% to 0.5% by weight of dry cement, such as nylon fibers. Crushed rubber may also be used, but it should be noted that the concentration of the additive in this case may be up to 5% by weight of dry cement. Cement mortars strengthened in this way show a significant increase in resistance to mechanical stresses and increase the service life of the cement concrete.

3.9.3 Radioactive Tracers

Radioactive compounds are sometimes added to cement slurries to make them easier to locate behind the casing. Initially, radioactive tracers were used quite extensively, but today safer cementing quality evaluation methods based on temperature changes in the borehole, acoustic signals, etc., have been developed.

Nevertheless, radioactive markers are still used in some secondary cementing operations. The principle of this technology is quite simple, before and after the cementing operations the radioactive logging is performed, the data obtained are compared and the location of plugging solution is determined by the change of radiation background. The most widespread radioactive substances used as tracers during well cementing are $_{53}I^{131}$ (half-life period is 8.1 days) and $_{77}Ir^{192}$ (half-life period is 74 days).

3.9.4 Mud Decontamination

Despite the undesirability of mixing drilling and cement slurries, in practice this phenomenon cannot be completely avoided. Some chemicals in drilling fluids, such as lignins, starches, cellulose, and lignosulfonates, can significantly increase the setting time of cement slurries, if not cause significant deterioration of their physical and chemical properties. To minimize such effects when mixing cement and drilling fluids, paraformaldehyde or a mixture of paraformaldehyde and sodium chromate is used.

4

Special Cement Systems

4.1 Thixotropic Cement

Initially, the phenomenon of thixotropy (thixotropy) should be briefly explained. The Greek word for "thixotropy" means "change when touched" and the scientific definition is as follows: It is the ability of a substance to reduce viscosity (liquefy) under mechanical action and to increase viscosity (thicken) when at rest. Thixotropic cements have a high enough mobility in circulation but lose it quickly, forming a gel-like structure when circulation ceases. A resumption of circulation restores the fluidity of the slurry.

Typically, thixotropic suspensions are Bingham fluids, i.e. their rheological behavior is determined by two basic parameters: the initial yield stress and the plastic viscosity. Bingham liquids behave like solids, i.e. they do not flow at low shear stresses, the minimum required value of the shear stress to start the flow of the liquid is called the initial value of the yield stress. Once the minimum shear stress is reached, the Binghamian fluid behaves like a Newtonian fluid, i.e. viscosity is independent of strain rate. In the case of nonthixotropic fluids, the value of the yield stress remains constant, regardless of whether the shear rate is increased or decreased. This behavior is explained by the fact that no change in the structure of the liquid occurs over time, whereas in thixotropic liquids a gel structure is formed (e.g. in the case of cement) and that, over time, a force exceeding the initial yield strength is required to restore the yield strength. The difference between these two values is called the degree of thixotropy of the fluid.

The vast majority of applications of thixotropic cements are found in the following operations:

- Securing intervals with a low fracture gradient
- Repair of corroded or partially fractured casing
- Prevention of gas migration

Oil and Gas Well Cementing for Engineers, First Edition. Baghir A. Suleimanov, Elchin F. Veliyev, and Azizagha A. Aliyev.
© 2023 John Wiley & Sons Ltd. Published 2023 by John Wiley & Sons Ltd.

Thixotropic cements, because they quickly form a gel structure, reduce the hydrostatic pressure created in the formation and, if necessary, easily return to their flowing state. All operations that use thixotropic cements are based on this principle. But it should be noted that with each static-dynamic cycle, the values of gel strength and yield strength will steadily increase. In practice, this means that at some point, these values may reach values sufficiently high to exceed the technical capabilities of the pumps. Therefore, it is risky to switch off pumps completely when pumping such plugging systems, and the number of static-dynamic cycles should be strictly controlled and not increased unnecessarily.

Today, the use of thixotropic cements has decreased to a large extent, as a rule, they are successfully replaced by low-density cementing slurries. However, they are still quite widespread. Modern thixotropic cement systems are represented by the following basic compositions:

- **Thixotropic clay-based cement compositions.** These are cement compositions based on Portland cement containing water-soluble clays, most often bentonite. Bentonite concentration ranges from 0.05% to 2% by weight of dry cement, and mortar density ranges from 1.381 to 2.521 kg/m³, respectively.
- **Thixotropic cement compositions based on calcium sulfate.** Calcium sulfate hemihydrate is the most widely used material for the preparation of thixotropic cement mortars because of its applicability to most Portland cements. Depending on cement, the optimum concentration is between 8% and 12% by weight of dry cement. The addition of calcium sulfate half-hydrate increases the volume requirement of the mixing water, resulting in lower slurry density. For compatibility with additives controlling the fluid loss of cement slurry, 1.5% polyvinyl alcohol by weight of dry cement is used. In addition to thixotropy, these compositions also have high sulfate resistance and a high coefficient of volumetric expansion.
- **Thixotropic cement compositions based on a mixture of aluminum sulfate and iron(II) sulfate.** This mixture has been developed for use with Portland cements containing less than 5% C_3A. The reaction of aforementioned sulfates with calcium hydroxide in cement slurry is accompanied by the formation of gypsum followed by the formation of ettringite. Aluminum sulfate is a powerful accelerator of cement setting, in the absence of iron sulfate in cement slurry would inevitably lead to the formation of irreversible gel structure. Iron(II) sulfate being a weak retarder of cement setting inhibits aluminum sulfate, preserving thixotropy of plugging solution during the whole injection time.
- **Thixotropic cements based on cross-linked polymer systems.** Thixotropic cements can also be produced by adding water-soluble crosslinkable polymers and a cross-linking agent to cement slurries. Optimal combinations of polymer, crosslinking agent and cement depend on well conditions and require individual selection.

4.2 Expansive Cement

The presence of a good bond between the hardened cement and the casing/formation is a very important factor affecting the efficiency of the isolation operation. A poor bonding of the hardened cement limits oil flow rates, significantly reducing the efficiency of stimulation operations during the production phase. Cement bonding to the string or reservoir is compromised for a number of reasons, but the main sources are the following phenomena:

- Incomplete removal of drilling mud from the annulus
- Expansion or shrinkage of the casing string due to changing thermobaric conditions in the well
- Contamination of cement slurry with drilling mud or formation water

These conditions almost inevitably create microgaps at the cement/formation or casing interface. In essence, a microsized annular space is created, through which fluids flow freely. This problem is solved by the use of post-consolidation expanding cement slurries. Due to the fact that the mortar is limited to the space between the formation and the column, cement expansion leads to compaction (i.e. a reduction in porosity of the cement stone) and, consequently, to an improvement in bonding quality. However, it should be noted that expanding cements do not solve the problem of cementing slurry contamination and maximum removal of drilling mud remains the most priority and effective method of improving cementing slurry bonding. The following main types of expansive grouts are distinguished:

1) **The cementing systems based on ettringite** are some of the most expensive cementing slurries in the oil industry. The mechanism of action is based on the formation of ettringite crystals in the cement slurry, which have a larger volume in comparison to the components from which they are formed. Currently, in the market, there are four types of expanding cement slurries based on ettringite formation:
    a) **Class K cement.** This cement is a mixture of Portland cement, calcium sulfate, lime, and anhydrous calcium sulphoaluminate. This cement consists of two separately burnt clinkers grinded together. K-type cement systems typically expand by 0.05–0.20%.
    b) **Class M cement.** This cement is a mixture of Portland cement and refractory cement based on calcium aluminate and calcium sulfate, or a mixture of Portland cement clinker, aluminate cement clinker, and calcium sulfate.
    c) **Class S cement.** This cement is a mixture of Portland cement high in C_3A and 10.5% to 15% gypsum.
    d) **Ettringite-based cement mortar** is the most common type of expanding cement and is produced by adding calcium sulfate hemihydrate to Portland

cement containing at least 5% C₃A. The main limitation of ettringite-based slurries is their inability to provide significant expansion at temperatures above 76 °C, due to their instability at high temperatures and decomposition to denser calcium sulphoaluminate hydrate and gypsum.

$$Ca_6Al_2(SO_4)_3(OH)_{12} \cdot 26H_2O \rightarrow 3CaO \cdot Al_2O_3 \cdot CaSO_4 \cdot 12H_2O \\ + 2(CaSO_4 \cdot 2H_2O) + 16H_2O$$

2) **Salt-cement solutions,** being one of the first expanding plugging systems used, are cement solutions containing high concentrations (up to 18%) of NaCl or Na_2SO_4. Cement slurry expansion mechanism is based on an increase in pore pressure as a result of the salt crystallization reaction. The expansion of cement-salt mortars is up to 0.35%. The application temperature can reach up to 200 °C.
3) **Cement mortars with the addition of finely ground aluminum.** Finely ground aluminum reacts with the alkalis in the cement slurry to form tiny hydrogen bubbles. However, this system is only effective when used in shallow wells, where the pressure created by the resulting gas bubbles exceeds formation pressure. The rate and degree of cement expansion are highly dependent on reservoir temperature and pressure, the degree of grinding and the concentration of aluminum particles, necessitating careful laboratory testing when selecting a cement composition.
4) **Cement mortars with the addition of calcined magnesium oxide.** The hydration reaction of magnesium oxide produces magnesium hydroxide which has a higher volume than the constituents, thus ensuring the expansion of the cement slurry. Cement systems containing calcined magnesium oxide offer average expansion values in the range of 1–1.5% at temperatures up to 280 °C. However, at temperatures below 60 °C, the hydration reaction is too slow and of no practical interest. The concentration of the additive in the cement mortar varies from 0.25% to 1.00% by weight of dry cement, depending on the well temperature.

4.3 Freeze-Protected Cement

Today, the permafrost zone covers approximately 25% of the planet's land area. Formed during an ice age, they are characterized by ground frozen to a depth of 600 m and are located mostly in Siberia and northern North America. Given the area's abundance of hydrocarbon deposits, well cementing operations in these environments are not uncommon. The cementing process in these environments has unique characteristics. For example, it is critically important not to allow the ground to thaw during drilling and completing as this may lead to subsidence of the melted earth which may lead to a breach of the well structure or other negative consequences. It is this condition that dictates the low hydration temperature of the cement slurry used, accompanied by the formation of a sufficiently strong cement stone at temperatures down to $-3\,°C$. Another peculiarity of the cementing process in these conditions is cementing of practically all casing from bottom hole to mouth, otherwise water released from the cement slurry due to expansion during freezing will cause various damages to the casing.

Slurries based on ordinary Portland cement are not suitable for such conditions as they freeze before sufficiently strong cement stone is formed. The most common cementing materials in permafrost conditions are:

- **Calcium aluminate cements** are heat-resistant cement compositions with a high aluminum or aluminum oxide content, designed for use in high-temperature wells. However, due to their rapid strength development and short setting time at near-zero temperatures, these cement systems have also proven highly effective in permafrost conditions.
- **Gypsum-Portland cement mix.** The gypsum sets quickly, gaining strength even at subzero temperatures, preventing the Portland cement from freezing. These mixes usually additionally contain sodium chloride as a freeze retarder. These cement systems generate less heat during hydration compared to calcium aluminate-based cements, making them practically indispensable for anchoring weakly cemented rocks at subzero temperatures.
- **Ultrafine Portland cements** are Portland cements with a Blaine fineness of more than $10\,000\,cm^2/g$, approximately three times that of normal Portland cement. The large specific surface area of these cements ensures early setting and hardening times.

4.4 Salt-Cement Systems

Salt-cement systems are cement systems containing significant quantities of salt (NaCl/KCl). The use of salt as an additive to cement is quite common for a number of reasons:

- Salts are naturally present in the mixing water in varying concentrations.
- When cementing well intervals pass through massive salt formations or clay interlayers, the use of salt as an additive avoids formation dissolution and clay swelling.
- In terms of affordability and accessibility, it is perhaps the cheapest and most common addition to cement slurries that effectively changes their physical and chemical properties.

As mentioned here, the use of mineralized water as mixing water is not uncommon and in some cases, e.g. in offshore mines, is unavoidable. Although this leads to changes in the physicochemical properties of cement slurries, such changes are considered acceptable due to the controllability of their effects. Nevertheless, the petroleum engineer must take into account the following features of seawater application in cement slurry preparation:

- shorter thickening time
- higher values of fluid loss of cement slurry
- higher values of cement strength at low temperatures during the initial stage of setting (first 24 hours)
- a slight decrease in cementing slurry viscosity
- better adhesion of cement to the casing
- increased tendency to foaming during mixing
- decrease of bentonite effect as a extender (to even this effect, bentonite needs to be prehydrated or even replaced by attapulgite).

4.5 Latex-Cement Systems

Latex is the common name for emulsions of dispersed polymer particles in an aqueous solution, appearing like milk. The size of dispersed polymer particles varies between 200 and 500 nm, the proportion of solids can be up to 50%. To improve the stability of the emulsion, surfactants are often added when mixing with Portland cement.

The use of latexes in Portland cement mortar dates back to the beginning of the twentieth century since clear advantages of this additive were already known at that time, such as

- improved pumpability of slurry
- decreasing permeability of cement stone
- increased tensile strength
- shrinkage reduction
- increased elasticity
- improvement of adhesion at the cement/steel and cement/rock interface.

However, in spite of the obvious positive effect of latex as a cement additive, the use of these mortars in plugging operations started much later in the 1960s of the twentieth century. In cementing operations, additional benefits of latex additives were discovered, such as:

- improved adhesion of cement to wet and oily surfaces
- reduced fluid loss of cement slurry

- higher resistance to negative effects of borehole fluids
- more resistant to perforation of the cement rock structure

The addition of latex to the mixing water has almost no effect on its appearance and consistency, but nevertheless results in a 20–35% reduction in water content due to the presence of solids in the latex.

For a long time, polyvinyl acetate-based systems have been used as latex, but the ineffectiveness of the application at reservoir temperatures below 50 °C has limited the wide application. A breakthrough has been the use of latexes based on styrene-butadiene to prevent gas migration in the annulus.

4.6 Corrosion-Resistant Cement

Hardened Portland cement is a relatively strong material, but the environment or the fluids injected into the well may still have a negative impact on the structure of the cement stone. The stability and durability of cement stone are extremely important when cementing wells designed for chemical waste disposal or for carbon dioxide injection (both for storage and enhanced recovery).

At cementing wells used for chemical waste disposal, the main danger is possible leakage caused by cement ring leakage or corrosion of metal surfaces. The issue of corrosion resistance of metal surfaces is solved mainly by processing them with different compositions based on polyesters and epoxy fibers. An alternative is the use of corrosion-resistant alloys such as Carpenter 20, Incoloy 825, and Hastalloy G, but due to their high cost, this approach is not often used.

Corrosion resistance of the cement ring is also a rather complicated and complex task aimed at modification of the composition of cement slurry depending on the chemical nature of the deposited materials. Thus, when dumping wastes containing weak organic acids, sewage water, or alkaline solutions, pozzolans, dispersants, and latex are added to cement slurry, or cement with a certain particle distribution is used. All the methods described here are essentially aimed at reducing the permeability of the cement stone in order to prevent or slow down the penetration of external fluids, which, in turn, leads to a reduction or prevention of corrosion processes.

When strong inorganic acids (sulfuric, hydrochloric and nitric acids) not compatible with Portland cement are deposited, organic polymeric cements, usually based on epoxy resin, are used to ensure sufficient corrosion resistance. These systems are also known as "synthetic cements". Organic polymer cements are produced by mixing epoxy resin with various hardeners. It is not uncommon to use solid fillers in such systems, to increase the density of the mortar, to reduce its cost and as a heat-absorbing element for the exothermic reaction occurring during the setting of the cement. The most common application for this purpose is silica flours. Organic polymer cements, due to their high corrosion resistance, are compatible with strong acids and bases (up to 37% HCl, 60% H_2SO_4, and 50% NaOH) at temperatures up to 93 °C. It is also characterized by higher strength values of the cement stone compared to conventional Portland cement.

Equally dangerous to the integrity of the cement stone is the presence of carbon dioxide, which leads to the leaching of the cement material from the cement matrix and a consequent increase in porosity and permeability values. This type of corrosion cannot be prevented. A simple solution to this problem would be to use synthetic cement, but unfortunately, such systems are not economically feasible for most CO_2 injection or sequestration projects.

To date, the cementing strategy for such wells has been limited to slowing corrosion rates as much as possible by reducing water-cement ratios, using fine cement, and adding latex and pozzolanic materials.

4.7 BFS Systems

Blast Furnace Slag (BFS) systems have been used for well cementing since the late 1960s. These systems are based on granulated blast furnace slag, an effective cementing material on its own. Although the addition of blast furnace slag to drilling fluids gives them cementitious properties, this practice has not become widespread.

BFS is a byproduct of steelmaking, produced by rapidly cooling molten (liquid) slag from a blast furnace with water. Depending on the source of the iron ore, there may be slight differences in the chemical composition of the slag. However, in order to ensure complete hydration, the fineness of crushed slag by Blaine should be at least 4000–5500 cm^2/g, regardless of the chemical composition. Unlike Portland cement, which sets when water is added, slag requires chemical activation to set. Chemical activators include compounds such as caustic soda, soda ash, Portland cement, lime, sodium sulfate/sodium silicate, or mixtures thereof.

When blast furnace slag is used in combination with Portland cement, no special chemical activators are required, as mentioned earlier, as two slag hydration activators are already present – gypsum and portlandite. Granulated blast furnace slag is mixed or ground together with Portland cement or a clinker-gypsum mixture in various proportions. As a rule, the proportion of slag in such a mixture is not more than 50%, but in some mixtures, it can be more than 80%.

The main advantages of BFS compared to Portland cement are

- higher sulfate resistance
- slower diffusion of chloride and alkaline ions through the cement stone
- lower permeability of cement stone

These advantages of BFS are associated with a reduction in the pore size of the cement stone with the addition of slag. A 30% reduction in porosity at 75% slag content is possible. The mechanism of this phenomenon is explained by the absence or reduction of large portlandite crystals with decreasing mass fraction of Portland cement in the mixture.

Slag cements are also widely used in the following well cementing technologies:

- Pressure cementing with the use of microfine cements
- In the use of salt-saturated cement slurries
- With CO_2 stable cements
- When using foamed cements
- Improving the performance of construction cements for use in well cementing operations

4.8 Engineered Particle-Size Distribution Cements

The modification of cement slurry properties is a complex task, depending on many factors such as solid phase content (including cement), presence of additives in the composition, amount, and type of mixing water, thermobaric conditions, etc.

This process is particularly difficult when working with slurries of high density (over 2100 kg/m^3) or conversely low density (under 1680 kg/m^3). In the first case, it is difficult to ensure the pumpability and stability of the slurry, while in the second case, the necessary strength of the cement stone is ensured. Traditionally, such complications are solved by changing the water-cement ratio, and adding various additives to cement slurries in the form of density reducers, viscosity modifiers, etc. However, another approach is possible – the use of cement mixtures with adjustable particle sizes. The essence of the method is to increase the packing density of cement solids leading to a reduction in the water-cement ratio and thus increasing the strength of the cement stone.

The packing density of particles is a geometric phenomenon and depends on the size and shape of the particles. The highest packing density of particles of the same size and spherical shape is 74%, but if a material consists of spherical particles with different diameters, the packing density can theoretically be increased to 95% and more. Such high-density values can be achieved by selecting the particle size distribution so that the voids between the larger particles are filled with smaller ones. Of course, in practice, it is practically impossible to achieve such packing density for a number of reasons, but the two main ones are the following:

- Mathematical models, in which calculations are made, do not take into account real shapes of particles (note that in reality, they are not perfectly spherical as assumed in the calculations).

Figure 4.1 Arrangement of cement particles in mathematical models.

- In mathematical models, such an arrangement of particles in space is assumed, at which a ratio of average diameters of two neighbor particles fluctuates within 7–10; in practice, such an arrangement of particles is certainly impossible (Figure 4.1).

However, although it is practically impossible to achieve maximum values for the packing density of cement particles, even achievable values have a significant effect on the properties of the cement slurry and stone listed here:

- Thickening time decreases with increasing particle-packing density
- Compressive strength increases with increasing packing density
- Slurry stability increases with increasing density of particle packing
- Fluid loss decreases with an increase in the packing density of particles
- Viscosity of the solution increases with the packing density of the particles

Thus, these parameters described in these slurries are less dependent on density values of slurries as compared to conventional cement systems. It is due to this feature that cements with adjustable particle size distribution have found wide application in low- and high-density cement slurries.

- **Low-density cement slurries.** For the preparation of such systems, extenders are used which have the side effect of increasing the porosity of the cement stone and the water-cement ratio. This, in turn, leads to a reduction in the strength of the cement stone and the stability of the slurry. On the contrary, the use of cements with controlled particle distribution ensures early setting of cement strength due to lower porosity values in comparison to traditional cementing systems used for this purpose. The stability of the slurry and, in the presence of inert fillers in the mixture, the chemical resistance of the setting cement are also considerably improved.

- **High-density cement slurries.** This type of slurry is traditionally prepared by adding weighting agents and reducing the water-cement ratio. As a result, the pumpability of the slurry is significantly reduced, making the use of viscosity modifiers practically necessary. All of these lead to excessive cement content in the slurry, which, in turn, has a negative effect on its stability as a whole. However, higher values of particle packing density of cements with adjustable particle size enable to increase cement content in cement slurry at low viscosity values.

4.9 Low-Density Cements

In well construction, it is not uncommon to encounter formations with low fracture gradients, unable to withstand even the hydrostatic pressure of the water column. Cementing such wells with conventional Portland cement is impossible, as the density of cement slurry is much higher than that of water. It is for this purpose that cementing slurries of reduced density are used. In practice, the upper limit of such systems' density is taken, as a rule, a bit higher than water density and makes 1200 kg/m^3.

As of today, the following three groups of cementing systems of this class have become the most widespread:

- cement slurries with extenders
- cement slurries with adjustable particle size distribution
- foamed cement

Since the first two groups of grouts have already been described, only the main features and characteristics of aerated cements are discussed here.

4.9.1 Foamed Cement

Foamed cement has been known in civil engineering for quite a long time, but has only been used for well cementing since the early 1980s of the twentieth century. These systems are coarse dispersions of Portland cement, gas (usually nitrogen), foaming surfactant, and various stabilizing additives. The density of the final slurry is controlled by changing the gas concentration. Among cementing slurries of low density, compositions of this group have the lowest cost. However, one should note that in contrast to other two groups of compositions for the preparation of foam cement, it is necessary to install special equipment for the injection of gas, which is quite difficult to provide in conditions of offshore fields. The undoubted advantage of using foam cement while maintaining all the necessary performance characteristics is the ability to reduce the density of the cementing solution to 420 kg/m³, which cannot be achieved either by using extenders or cements with adjustable grain-size composition.

In addition to low density, foamed cements also offer a number of advantages:

- Relatively high compressive strength developed in a reasonable time
- Low likelihood of formation damage factor for water-sensitive formations
- Low likelihood of annular gas exchange
- The possibility of cementing formations with a high probability of fluid loss, as the density of the system can be adjusted during the cementing process by simply changing the gas concentration.

The main factor influencing the efficiency of the foamed cement slurries used is the stability of the resulting compositions. The stability of aerated systems, in turn, depends on the following factors:

- Foaming agent
- Gas amount
- Chemical composition of the solution
- Thermodynamic factors
- Mixing methods and conditions

Stable foam cements have a spherical, discrete, disconnected pore structure with a distinct cement matrix. In contrast, unstable foamed cements have non-spherical and interconnected pores formed by the rupture and coalescence of gas bubbles, and as a consequence, this significantly impairs their isolation properties.

4.10 Flexible Cement

The mechanical properties of the hardened cement stone are extremely important to ensure safe and efficient well operation. Today, when designing cementing operations, not only direct axial loads on cement sheath but also such parameters as Poisson's coefficient and Young's modulus are taken into account. Elasticity of cement slurries allows to avoid or significantly decrease negative consequences of such detrimental processes for the integrity of annular space isolation as:

- Perforation and hydraulic fracturing
- Thermodynamic processes, occurring in cement slurry. For example, expansion of bound water in cement matrix with the increase in temperature
- Tectonic processes

Improving the elasticity of cementing materials is achieved by the following main methods:

- **Reducing the density of the mortar.** Lightening additives additionally absorbing water (i.e. sodium silicate and bentonite) increase its total content in a slurry, increasing cement stone elasticity. However, it should be noted that this method negatively affects such properties of cement stone as compressive strength and permeability.
- **The addition of elastic particles into the cement slurry.** This method is borrowed from civil engineering, where the addition of rubber particles to improve the quality of road surfaces, noise insulation, giving anti-impact properties, etc., has been used for quite a long time. In well construction, this method has been successfully used to secure the production casing bottom hole, which considerably reduces the negative effect of perforation on annulus integrity. There are also cementing systems with adjustable particle sizes in which one of the fractions is represented by elastic particles. Thermoplastics such as polyamide, polypropylene, and polyethylene or polymers such as styrene divinylbenzene or styrene-butadiene are used as elastic particles. These compounds, due to their low density, also reduce the density of the cement slurry.
- **Application of elastomeric composites.** As a rule, these are composite materials based on rubber with various fillers to regulate the mechanical properties.
- **Application of fiber additives.** This method has also been used in civil engineering for quite a long time. In the construction of wells usually as fiber additives are used nylon fibers or metal microfiber.

4.11 Microfine Cements

The fine particle size (4–15 μm) in cement mixtures of ultrafine grinding provides a high density of the specific surface of the mixture (more than $1000\,m^2/kg$) and as a consequence a high reactivity. The most common mixtures are based on Portland cement, although there are also slag-cement compositions. The advantage of such systems is their improved filtration characteristics. Due to the high reactivity, these mixtures often include various setting retarders. The main applications are as follows:

- Squeeze cementing
- Sealing of casing leaks
- Permafrost cementing
- Cementation of upper intervals in deep wells.

4.12 Acid-Soluble Cements

The control of fluid loss, both in drilling and cementing wells, is a fairly common practice. As a rule, for this purpose, injection of various thixotropic cements, fibrous and flake materials are used. Due to the fact that the injected agents remain in the formation, the earlier described methods are not suitable if the intake interval is in the productive zone of the formation and there is a need for materials that can be easily removed after the completion of the well. Such a temporary isolation treatment based on magnesium oxychloride (Sorel cement), obtained by mixing powdered magnesium oxide with concentrated magnesium chloride solution, was developed in the early 1990s. Despite the high strength of the resulting cement stone, it dissolves easily when exposed to acid. These cements are well compatible with standard cement additives and can be modified for a wide range of reservoir conditions. In addition to controlling fluid loss, acid-soluble cements are also widely used for temporary isolation of various reservoir intervals.

4.13 Chemically Bonded Phosphate Ceramics

The peculiarity of these systems is that the mechanism of cement stone formation is based on the process of synthesis of phosphate compounds. Solidification of phosphate cements occurs by the interaction of powdered components of cementing mixture with the mixing fluid. This fluid is the phosphate-containing compounds (aqueous solutions of phosphoric acids and acid phosphate solutions). Powdered components of cementing mixture, as a rule, contain oxides and hydroxides of different metals, glass, salts, etc. The chemical mechanism of the hardening of this mixture is based on acid–base interaction of the mixing fluid and the powdered components of the mixture.

For phosphate ceramics, curing is due to the chemical interaction of the initial solid powdered component with the mixing fluid containing phosphate anions. Thermal orthophosphoric acid solutions of 40–70% concentration, salts of orthophosphoric acid, as well as aluminophosphate, aluminochrome phosphate, magnesium phosphate, chrome phosphate binder, are most widely used as a hardener.

Calcium aluminate-based phosphate ceramics are used for cementing geothermal wells. Cements based on magnesium-potassium phosphates developed for storage and encapsulation of radioactive waste have also found application as fast hardening cements for cementing operations. These systems are compatible with a number of cement additives and can also be used in a fairly wide range of reservoir conditions. It should be noted that phosphate ceramics have significantly higher compressive strength in comparison with Portland cement.

4.14 Special Cement Systems

4.14.1 Nonaqueous Cement Systems

Nonaqueous cementing slurries have been used for quite a long time for repair and insulation works or for water shut-off in the process of drilling. The peculiarity of cementing solutions of this group is the absence of water in the injected composition, thanks to which the process of cementing does not start before the solution comes into contact with formation water. Thus, the probability of the clogging of oil-saturated reservoir sections is practically minimized and isolation of highly watered areas and channels is provided. As a rule, microdispersed cement is used to ensure good filtration characteristics of cementing slurry. As a mixing fluid, it is possible to use both fluids on the hydrocarbon base (for example, diesel fuel) and more complex compositions, which include surface-active agents and various solvents. However, it should be noted that the use of hydrocarbon-based fluids as a mixing fluid has a number of drawbacks, the most significant of which is the formation of a hydrophobic film on the cement particles. The resulting hydrophobic film serves as a kind of barrier to the reaction of hydration substantially, slowing it down, that is, the process of thickening and hardening of the cement even in contact with formation water is strongly prolonged.

4.14.2 Storable Cement Slurries

The preparation of cement mix at a well is a complex process that requires additional equipment and personnel. Often it also requires additional storage space for cement and cement additives, and the mineralization of the mixing water is dictated by the nearest available source. Thus, every cementing project is initially subject to a number of constraints that are not changeable. The aforementioned factors can be eliminated when using readymade cement slurry, i.e. when liquid cement slurry is delivered to the wellbore. Such technology has started to develop since the early 1990s of the twentieth century. The storage time of cement slurries of this group is almost unlimited. At the same time, these solutions can be made practically from any class of Portland cement and are compatible with many basic cement additives. As a rule, acids serve as a retarder of setting, and at pumping into a well, the slurry is additionally added as an activator, starting the process of hydration. Sodium silicate is most commonly used for this purpose.

5

Cementing Equipment

5.1 Surface Equipment

A simplified schematic illustration of the cement slurry preparation, transportation, and storage process is shown in Figure 5.1.

Initially, cement is delivered to the central warehouse, usually stored in bags, but storage in dry cement **silos** is also possible (Figure 5.2). The standard design of a silo is a metal cylinder, closed at the top with a lid with ventilation holes and filters, ending at the bottom with a cone with a hole and a gate valve installed in it for cement discharge. Installed vertically on supports. Loading systems are available for transferring cement from one silo to another or to a mixer, cement carrier, or supply vessel. With a pneumatic transfer system, multiple silos are connected on a continuous basis to save time and labor. In humid climates, a dehumidifier can additionally be installed in the system.

Loading and unloading equipment usually also include various pneumatic discharge devices (Figures 5.3 and 5.4), mechanical screws or combined systems (for unloading from ships).

Cement and dry additives are, in most cases, mixed in pneumatic mixing tanks (10–20 tons capacity) directly in the central warehouse and the ready-to-use mixes are delivered to the well (Figure 5.5). Bulk solids are usually pneumatically pumped into the mixing tanks, and packaged materials are poured through a hopper located on top. The frame of the mixing tank also incorporates weight detection equipment. The materials inside the tank are mixed by an airflow injected through nozzles at a pressure of 2.5 bar. Typically, this operation is repeated several times to ensure that the required volume of cementing mixture is achieved.

Oil and Gas Well Cementing for Engineers, First Edition. Baghir A. Suleimanov, Elchin F. Veliyev, and Azizagha A. Aliyev.
© 2023 John Wiley & Sons Ltd. Published 2023 by John Wiley & Sons Ltd.

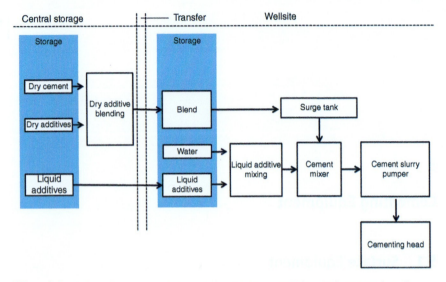

Figure 5.1 Schematic representation of storage, transportation, and preparation of cement slurry.

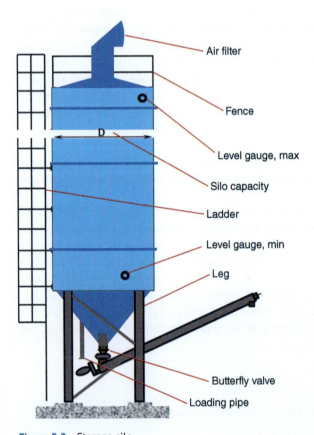

Figure 5.2 Storage silo.

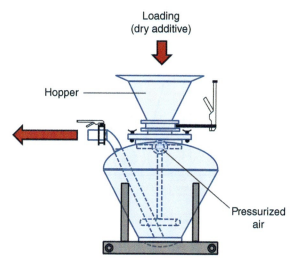

Figure 5.3 Pneumatic loading (dry material).

Figure 5.4 Screw-type unloader.

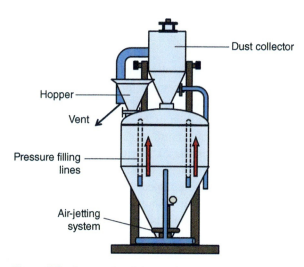

Figure 5.5 Pneumatic mixing tank.

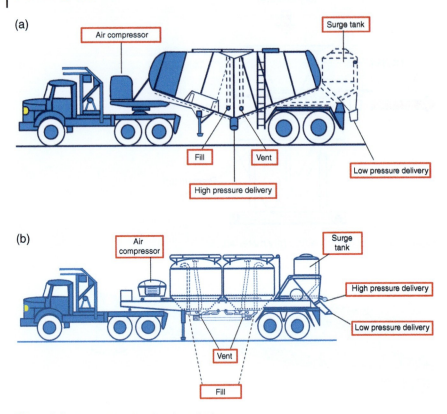

Figure 5.6 Cement trucks of various designs.

Depending on the logistical features of the well, different vehicles are used to transport the cement mixture: cement trucks (Figure 5.6), ships (offshore fields), or helicopters (hard-to-reach regions).

Sometimes cement mixture is delivered to the well before the work is done, or pure cement without additives is delivered and cement mixture is prepared on site. In such cases, it is required to provide conditions for cement storage at the site. To this end, cement silos similar to those used in the central warehouse, but of a smaller size, are used. These silos are divided into two large groups: operated under pressure and without pressure (atmospheric). In silos of the first group, airflow under pressure of 3 bar is used for movement of cement mixture inside the silo, and in the second group, airflow pressure pumped through the porous silo floor is insignificant and amounts to about 0.2 bar. It should be noted that the silos operated under pressure can be used in both vertical and horizontal positions, while the silos of the second group are operated in a vertical position only. All silos

5.1 Surface Equipment

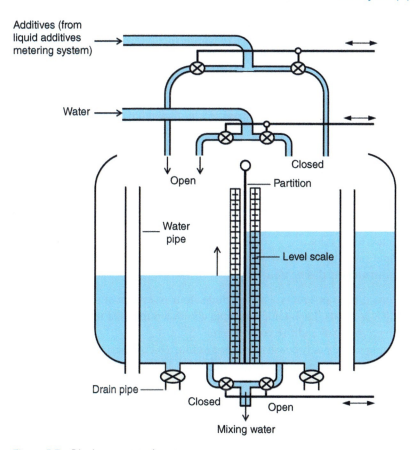

Figure 5.7 Displacement tank system.

used directly on the well are equipped with a set of supporting legs for installation on an uneven surface and loading on trailers.

The volume of mixing fluid is measured in measuring tanks of 10 or 20 barrels. Measuring tanks are also used to control the volume of displacement fluid (Figure 5.7).

Cement additives added to the mixing fluid can be mixed with it either before or after it is pumped into the measuring tank. In the second case, however, an additional fluid dosing system will be required.

There are two basic approaches to preparing a batch fluid:

- **Premixing liquid additives with water.** In this method, a premixed mixing fluid is delivered to the wellbore.
- **On-the-fly mixing**. In this process, the mixing fluid is prepared down hole.

5 Cementing Equipment

Premixing of liquid additives with water is the simplest way to prepare cementing fluid, but it has a number of disadvantages, the most important of which are:

- The need for additional tanks at the well site, which is not always possible especially in offshore fields.
- If cementing operations are suspended, it is unlikely that this fluid will be used in another well.
- It is practically impossible to provide an additional volume of cementing fluid in case of need.

Due to the aforementioned disadvantages of using a premixed mixing fluid, the second group technologies are preferable. On-the-fly mixing methods use a semiautomatic or automatic dosing system that feeds the correct amount of additives into the measuring tank.

All liquid additive dosing systems consist of two main parts – the storage/transfer unit and the dosing unit.

- The storage/transfer unit typically includes four reservoirs of varying volumes from 1000 to 4000 l. Each tank is equipped with a pump and mixing system to avoid segregation of additive components.
- The dosing unit usually consists of three or four 10-l measuring tanks with visible level scales, where a premeasured amount of additives is poured (Figure 5.8). When preparing the mixing fluid, these additives are poured into one of the two

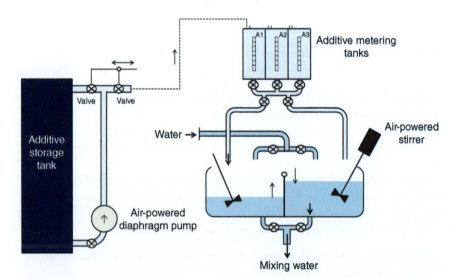

Figure 5.8 Liquid additive metering system.

sections of the measuring tank filled with water. Since the volume of the measuring tank usually does not exceed 20 barrels (3179 l), if a larger volume of mixing fluid is needed, this operation is repeated. It should be noted that this process can be automated and there are systems without measuring tanks based on the automation of chemical feeding with an accuracy of ±2%.

For smooth operation of cement slurry preparation system, it is necessary to ensure constant pressure in the mixer bowl and cement (or cement mixture) delivery rate. The second equally important factor is the prevention of pulsating flow during cement feeding into the mixer. For this purpose, equalizing tanks are used, which are cylindrical tanks of up to 2000 l in volume, which connect the mixer and the injection line (Figure 5.9). This device maintains the pressure above the mixer bowl and the hydrostatic head of the material.

Different types of mixers are used for mixing cement slurry, which are devices in which the mixing fluid flow is mixed with the cement slurry flow in specified proportions depending on the required output volume. The most common types of mixers are described as follows.

Hydraulic mixer is an ejector-type mixer consisting of a feed hopper, mixing chamber, discharge nozzle, and storage tank for cement slurry (Figure 5.10). Maximal productivity of such a pump is estimated in flow rate of dry material (cementing mixture) a little bit more than 1 ton/min.

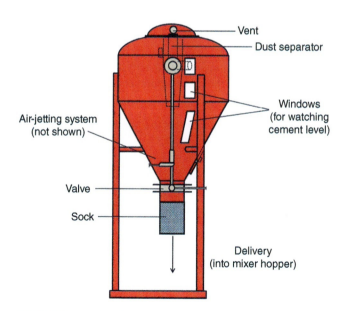

Figure 5.9 Surge tank.

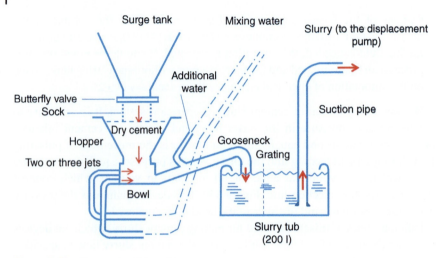

Figure 5.10 Conventional jet mixer.

Initially, the cementing mixture poured into a hopper simultaneously with the mixing fluid flow is fed into the mixing chamber; high pressure of mixing fluid jets mixes the slurry forming turbulent flow in a casing. It is due to the turbulence of the flow in the casing that the slurry components are thoroughly mixed and formed. From the casing, the slurry flows to the storage tank for density control, where the required density of the cement slurry is achieved by changing the water-cement ratio.

Recirculation hydraulic mixer. The maximum capacity of this type of mixer is just over 2 tons/min. The following are the main differences from conventional hydraulic mixers (Figure 5.11):

- There is a remote-controlled sliding flap between the hopper and the mixing chamber.
- The density of the slurry is controlled by a remote-controlled sliding gate.
- The spray liquid is displaced from the storage tank by a recirculation stream supplied by the centrifugal pump. The centrifugal pump also feeds the displacement pumps and recirculates part of the solution through the mixing system.

This method improves rheological indicators and homogeneity of cement slurry, making the process of density control much easier.

It should be noted that there is a sufficiently large variety of mixers that are not of the ejector type. Average productivity of these mixers is usually close to 2 tons/min.

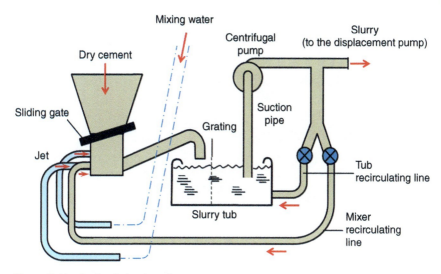

Figure 5.11 Recirculating jet mixer.

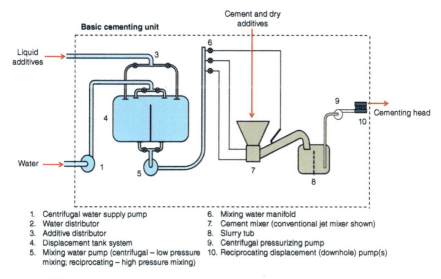

1. Centrifugal water supply pump
2. Water distributor
3. Additive distributor
4. Displacement tank system
5. Mixing water pump (centrifugal – low pressure mixing; reciprocating – high pressure mixing)
6. Mixing water manifold
7. Cement mixer (conventional jet mixer shown)
8. Slurry tub
9. Centrifugal pressurizing pump
10. Reciprocating displacement (downhole) pump(s)

Figure 5.12 Mixing and pumping equipment on rig site (typical setup).

From a storage tank, cement slurry is fed to high-pressure pumps and pumped into the borehole through a cementing head (Figure 5.12).

Today, it is quite common to use universal cementing complexes that combine all the aforementioned equipment units or their functions on different chassis (Figure 5.13).

Figure 5.13 Universal cementing unit (Halliburton Corporation).

5.2 Casing Types

The cementing job design starts at the bottom hole (i.e. production casing) and runs backward to the wellhead (i.e. to the conductor). The number and type of casing strings vary depending on available logistics, geological conditions, etc.

Casing is classified by technical characteristics such as outside/inner diameter, weight, and material. The choice of which type of casing is used depends on:

- The depth of the cemented interval
- The size of the hole to be cemented, i.e. the bit diameter, which is used when drilling the interval (Figure 5.14)
- The requirements of the well (i.e. yield strength, crush resistance, etc.)
- Reservoir and downhole pressure

Here are mentioned the main types of casing and their purpose.

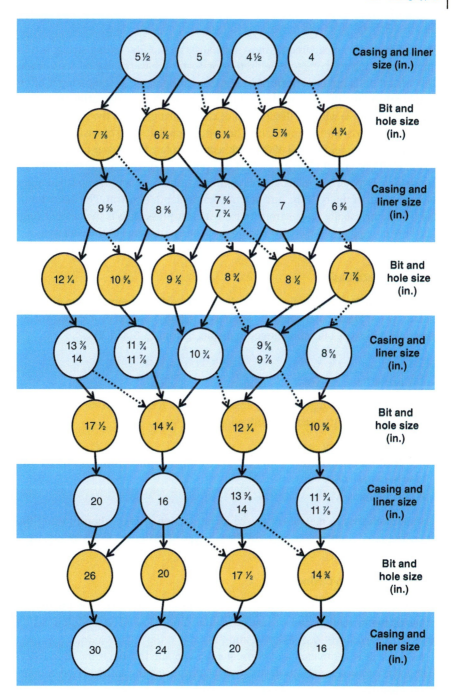

Figure 5.14 Possible casing diameter depending on the bit diameter used during drilling.

5.2.1 Conductor Casing

The conductor is considered by most specialists as the first casing string, but in some cases, such as in offshore fields, the conductor is lowered to the borehole pit. This operation is rarely performed by the cementing team, as the casing is often simply driven into the ground, due to the shallow depth of the lowering (not more than 10 m). That is why the description of casing string type in the literature predominantly starts with a conductor. The conductor is a string of casing designed to separate the upper interval of the rock section, isolate freshwater horizons from contamination, install blowout control equipment, and suspend subsequent casing strings. Depending on geological conditions, the conductor is installed at an average depth of up to 100 m, and at a maximum depth of 600 m. The diameter of the conductor usually ranges from 16 to 26 in. It is pressed, as is the cement ring. The bore hole pit and the conductor are mandatory elements of the well design.

5.2.2 Surface Casing

Surface casing is run to prevent the collapse of weakly cemented formations, to isolate groundwater and shallow aquifers, and to provide protection during blowouts. This type of casing ranges in diameter from 13-3/8 to 20 in.

5.2.3 Intermediate Casing

The intermediate casing is usually installed in the transition zone to abnormal reservoir pressures to isolate the following zones:

- Unstable sections of the well
- Zones of lost circulation
- Low pressure zones
- Production zones

Some such strings may also be used as production strings when running the liner. The most common casing sizes for this type of casing are 9-5/8 or 10¾ in.

5.2.4 Production Casing

The production casing is the last casing to be installed to isolate the productive zones of the formation and extract formation fluids. The most common casing sizes of this type are 4½, 5, 5¾, 6⅝, and 7 in.

As the well is completed and consequently exploited through the production string, it is exposed to various destructive factors during the entire period of operation, ensuring high quality of cementing works is extremely important.

5.2.5 Liner

A liner is a type of casing string that does not reach the wellhead (surface). The liner is installed in the previous casing string by means of a special overlapping suspension system at a distance of 50–65 m. An extremely important technical characteristic of liners is their ability to withstand collapse pressure. It is not uncommon for a well design where the liner and the previous string act as the production string. The use of liners has the following advantages:

- It reduces logistical and time expenditures for technical operations in the well. For instance: running and cementing of the production string.
- It allows completing with tubing of bigger diameter.

It must be noted that there are certain difficulties when using liners, such as joint leakage and small annular clearances between liner and bore hole.

According to their purpose, the liners can be divided into two groups – those used for drilling and for well cementing.

The liners used in drilling serve for isolation of zones of anomalous pressures.

The liners belonging to the second group, in turn, are also divided into the following four main types (Figure 5.15):

Production liner. Liners of this type perform the functions of a production casing.

Tie-back liner is a section of casing installed in the interval from the wellhead to the top of the liner located in the well. Reasons for installation of this type of liner are usually: insufficient strength properties of the well construction or failure of integrity of the intermediate casing.

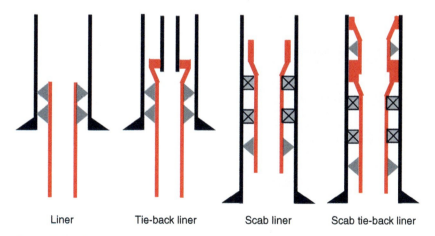

Figure 5.15 Types of liners.

Scab liner is a section of casing isolated by packers, which does not reach the wellhead and is installed for the repair of damages of intermediate casing string.

Scab tie-back liner is a casing section installed over the top of the liner in the well, which does not reach the wellhead.

5.3 Technical Characteristics of Casing

5.3.1 Steel Grades

The casing grade is a measure of the strength of the casing material and is governed by the API specifications (API Spec. 5CT "Specification for Casing and Tubing") and the International Organization for Standardization (ISO Spec. 11960 "Petroleum and Natural Gas Industries-Steel Pipes for Use as Casing or Tubing for Wells"). It is necessary to note that despite the fact that in former USSR countries, GOST standards are mainly used as industry standards for the production and testing of casing pipes, their difference with aforementioned standards is minimal. Steel casing pipe grades according to API 5CT/ISO 11960 are listed in Table 5.1.

Casing manufacturers use many different methods and technologies of steel processing, which, in turn, leads to high variability in the chemical composition of casing pipes and, as a consequence, their technical performance. In the API

Table 5.1 Casing steel grades according to API 5CT/ISO 11960.

Group	Grade	Yield strength (ksi) min	Yield strength (ksi) max	Tensile strength (ksi) min
1	H40	40	80	60
	J55	55	80	75
	K55	55	80	95
	N80	80	110	100
	R95	95	110	105
2	M65	65	85	85
	L80	80	95	95
	C90	90	105	100
	T95	95	110	105
3	P110	110	140	125
4	Q125	125	150	135

Figure 5.16 Interpretation of steel grades according to API standards.

classification, each steel grade is designated by a letter and a number. The letter refers to the chemical composition of the steel, and the number indicates the minimum yield strength in kilopound per square inch. For example, N-80 casing pipe has a minimum yield strength of 80 kilopounds per square inch (Figure 5.16).

As the minimum yield strength for casing increases, the risk of cracking increases in addition to mechanical strength under certain operating conditions. Casing with a yield strength greater than 140 ksi does not qualify for the API classification.

To avoid premature casing failure in sour environments, the National Association of Corrosion Engineers (NACE) has defined the following operating conditions for various casing grades (Table 5.2).

According to the API classification, casing steel grades are divided into four main groups (Table 5.3):

Group – 1. This group includes steels with relatively low strength grades **H, J, K, N**

- **H-40 grade steel** is not typically used in production tubing or production string because of its relatively low yield strength and small cost differential with respect to J-55 grade tubing
- **J-55 grade** is the most popular steel grade for tubing string production and is used in shallow wells up to 3000 m.
- **K-55 grade** is almost identical to J-55 grade, except that the minimum tensile strength of K-55 grade is higher.

Table 5.2 Casing steel grades for sour environments as recommended by the National Association of Corrosion Engineers.

Regardless of temperature	At temperature ≥ 65°C	At temperature ≥ 80°C	At temperature ≥ 107°C
H40	N80 тип Q	N80	Q125
J55	R95	P110	
K55	C100		
M65			
L80 type 1			
C90 type 1			
T95 type 1			

Table 5.3 Chemical composition of steel grades according to API classification for casing and tubing (wt%).

Group	Grade	Type	C min	C max	Mn min	Mn max	Mo min	Mo max	Cr min	Cr max	Ni max	Cu max	P max	S max	Si max
1	H40	—	—	—	—	—	—	—	—	—	—	—	0.03	0.03	—
	J55	—	—	—	—	—	—	—	—	—	—	—	0.03	0.03	—
	K55	—	—	—	—	—	—	—	—	—	—	—	0.03	0.03	—
	N80	1	—	—	—	—	—	—	—	—	—	—	0.03	0.03	—
	N80	Q	—	—	—	—	—	—	—	—	—	—	0.03	0.03	—
2	M65	—	—	—	—	—	—	—	—	—	—	—	0.03	0.03	—
	L80	1	—	0.43	—	1.90	—	—	—	—	0.25	0.35	0.03	0.03	0.45
	L80	9Cr	—	0.15	0.30	0.60	0.90	1.10	8.00	10.0	0.50	0.25	0.02	0.01	1.00
	L80	13Cr	0.15	0.22	0.25	1.00	—	—	12.0	14.0	0.50	0.25	0.02	0.01	1.00
	C90	1	—	0.35	—	1.20	0.25	0.85	—	1.50	0.99	—	0.02	0.01	—
	C90	2	—	0.50	—	1.90	—	—	—	—	0.99	—	0.03	0.01	—
	C95	—	—	0.45	—	1.90	—	—	—	—	—	—	0.03	0.03	0.45
	T95	1	—	0.35	—	1.20	025	0.85	0.40	1.50	0.99	—	0.02	0.01	—
	T95	2	—	0.50	—	1.90	—	—	—	—	0.99	—	0.03	0.01	—
3	P110	e	—	—	—	—	—	—	—	—	—	—	0.03	0.03	—
4	Q125	1	—	0.35	—	1.35	—	0.85	—	1.50	0.99	—	0.02	0.02	—
	Q125	2	—	0.35	—	1.00	—	—	—	—	0.99	—	0.02	0.02	—
	Q125	3	—	0.50	—	1.90	—	—	—	—	0.99	—	0.03	0.01	—
	Q125	4	—	0.50	—	1.90	—	—	—	—	0.99	—	0.03	0.02	—

- **N-80 grade steel** is one of the oldest and is suitable for manufacturing casing pipes. It is subdivided in turn into two subtypes: **N-80-1** and **N-80-Q**. However, N80Q has higher strength characteristics because it is produced by quenching and hardening, in contrast to N-80-1 steel produced by stabilizing and hardening.

Group – 2. This group includes steels with sulfur-resistant and corrosion-resistant characteristics of **L, C, and T grades**.

- **L-80 grade steel.** Subdivided into three types. **L80-1, L80-9CR, and L80-13Cr.** Steel grades L80-9CR and L80-13Cr have higher corrosion resistance, but are quite expensive and difficult to produce, so they are used only in highly corrosive environments.
- **Steel grade C-90** is a relatively new steel grade that is subdivided into two types: **C90-1** and **C90-2**. C90-1 is recommended for use in environments (at pH <7). Today this steel grade is produced to order and is gradually being replaced by T-95 grade.
- **T-95 steel grade** has high strength and resistance to sulfide stress corrosion cracking and is subdivided into two subtypes: **T95-1** and **T95-2**.

Group – 3. This group includes high-strength steel grade **P-110** allowing heat treatment by quenching and tempering.

Group – 4. This group includes high-strength steel grade **Q-125**.

5.3.2 Strength Characteristics of Casing

The strength characteristics of the casing are determined by the following basic parameters:

- The yield strength of the pipe and joint
- Burst pressure
- Collapse pressure

Yield strength according to API (API 5C3/ISO 10400) is the tensile stress required to increase the total length of a specimen by 0.5–0.7% depending on the grade of the tubing, calculated using the Formula (5.1):

$$T_y = \frac{\pi Y_p}{4}\left(D^2 - d^2\right) \qquad (5.1)$$

where Y_p – value of yield strength of pipe steel grade,

D – outside diameter of the casing,

d – inside diameter of the casing.

Burst pressure is defined by API (API 5C3/ISO 10400) as the maximum internal pressure required to cause the steel to flow and is calculated using Barlow Formula (5.2):

$$P_b = 0.875 \left(\frac{2Y_p t}{D} \right) \qquad (5.2)$$

where Y_p – value of yield strength of pipe steel grade,

D – outside diameter of the casing,
t – wall thickness of the casing.

In this case, the value of this parameter must be at least 0.875 of the regulated for the casing of this steel grade according to the specification of the API.

The collapse pressure (P_c) is defined in API 5C3/ISO 10400 as the maximum external pressure required to fracture the casing in the absence of axial load and internal pressure.

API Bulletin (API 5C3/ISO 10400) details the casing collapse pressure in four empirical formulas. Depending on the steel grade and the ratio of inner diameter to casing wall thickness, engineers use one of these formulas for practical calculations (Table 5.4).

1) **Elastic collapse P_c**

$$P_c = \frac{46.95 * 10^6}{\dfrac{D}{t}\left(\dfrac{D}{t} - 1\right)^2} \qquad (5.3)$$

where D is the outside diameter of the casing; t is the wall thickness of the casing.

2) **API transition collapse pressure P_{tr}**

$$P_{tr} = Y \left[\frac{F}{\left(\dfrac{D}{t}\right)} - G \right] \qquad (5.4)$$

where D is the outside diameter of the casing; t is the wall thickness of the casing; F, G are constants depending on the steel grade used; Y is the yield strength.

Table 5.4 Empirical formulas for determining the casing collapse pressure.

1. API elastic collapse pressure formula range

Grade	D/t
H-40	≥42.64
J-K-55	≥37.21
N-L-80	≥31.02
C-90	≥29.18
C-95	≥28.36
P-110	≥26.22
Q-125	≥24.46

$$P_c = \frac{46.95 \times 10^6}{\frac{D}{t}\left(\frac{D}{t}-1\right)^2}$$

2. API transition collapse pressure formula range

Grade	D/t	F	G
H-40	16.40–42.64	2.063	0.0325
J-K-55	14.81–37.21	1.989	0.036
N-L-80	13.38–31.02	1.998	0.0434
C-90	13.01–29.18	2.017	0.0466
C-95	12.85–28.36	2.029	0.0482
P-110	12.44–26.22	2.053	0.0515
Q-125	12.11–24.46	2.092	0.0565

$$P_{tr} = Y_p \left[\frac{F}{\left(\frac{D}{t}\right)} - G\right]$$

3. API yield collapse pressure formula range

Grade	D/t
H-40	≤16.40
J-K-55	≤14.81
N-L-80	≤13.38
C-90	≤13.01
C-95	≤12.85
P-110	≤12.44
Q-125	≤12.11

$$P_{yp} = 2Y_p \left[\frac{\frac{D}{t}-1}{\left(\frac{D}{t}\right)^2}\right]$$

4. API plastic collapse pressure formula range

Grade	D/t	A	B
H-40	16.40–27.01	2.95	0.0465
J-K-55	14.81–25.01	2.991	0.0541
N-L-80	13.38–22.47	3.071	0.0667
C-90	13.01–21.69	3.106	0.0718
C-95	12.85–21.33	3.124	0.0743
P-110	12.44–20.41	3.181	0.0819
Q-125	12.11–19.63	3.239	0.0895

$$P_P = Y_p \left[\frac{A}{\left(\frac{D}{t}\right)} - B\right] - C$$

3) **API plastic collapse pressure** P_P

$$P_P = Y\left[\frac{A}{\left(\frac{D}{t}\right)} - B\right] - C \qquad (5.5)$$

where D is the outside diameter of the casing; t is the wall thickness of the casing; A, B, and C are constants depending on the steel grade used; Y is the yield strength.

4) **API yield collapse pressure.** This parameter refers to thick-walled casing pipes. Despite the name, the mechanism of this type of collapse has nothing to do with the value of yield strength and is described by the Lamé equation, since, in some cases, in thick-walled pipes ($D/t < 15\pm$), the shear stress may exceed the yield strength of the pipe, leading to its collapse.

$$P_{YP} = 2Y\left[\frac{\frac{D}{t} - 1}{\left(\frac{D}{t}\right)^2}\right] \qquad (5.6)$$

where D is the outside diameter of the casing pipe; t is the wall thickness of the casing pipe; Y is the yield strength.

5.3.3 Weight Per Unit Length of Tube

The American Petroleum Institute (API) defines three types of pipe weights:

- **Nominal weight.** The notion of "nominal weight" is used when ordering casing pipe and design is based on calculating the weight of 1 m of pipe with threads and couplings. It is calculated by Formula (5.7):

$$W_n = 0.24(D-t)*t + 1.635*10^{-3}D^2 \qquad (5.7)$$

where D – outside diameter of the casing, mm; t – wall thickness of the casing, mm.

- **Weight of smooth pipe.** Weight of pipe without coupling and threads. Calculated by Formula (5.8):

$$W_{s.p.} = 0.24(D-t)t \qquad (5.8)$$

where D – outside diameter of the casing, mm; t – wall thickness of the casing, mm.

- **Weight of connection with thread and coupling.** The average weight of the pipe including the thread on both ends and the coupling on one end. Calculated by Formula (5.9):

$$W = \left[6.1 - 1.058 * 10^{-2}\left(N_c + 2J\right)\right]W_{s.p.} \tag{5.9}$$

where N_c – coupling length, m; J – distance from the end of the pipe to the center of the coupling at mechanical tightening, m.

5.3.4 Connection Types of Casing

Casing pipes are connected by threaded or welded connections. The welded type of connection is practically not used today. The threaded type, in turn, is subdivided into coupling (Figure 5.17) and noncoupling (Figure 5.18). The API distinguishes the following types of casing joints:

Connection with coupling (T&C – Threaded and Coupled)
- LT&C – long thread and coupled
- ST&C – short thread and coupled
- BTC buttress thread

Connection without coupling (Integral Connection)
- EU – external upset (Figure 5.19a)
- IU – internal upset (Figure 5.19b)
- NU – nonupset (Figure 5.19c)
- IEU – internal-external upset (Figure 5.19d)

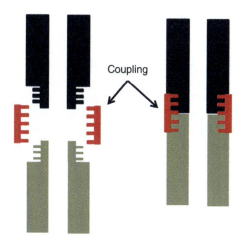

Figure 5.17 Connection with coupling.

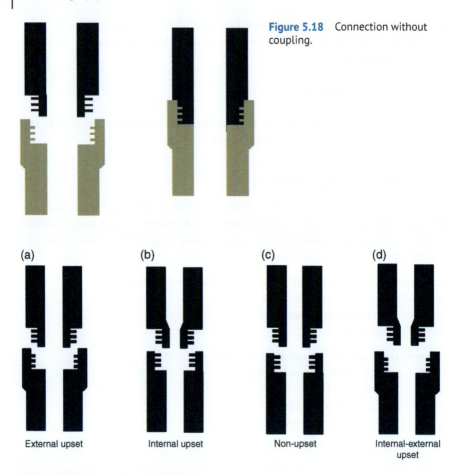

Figure 5.18 Connection without coupling.

Figure 5.19 Types of non-coupling connection.

5.4 Casing Hardware

Casing hardware includes a set of devices that are used to run and cement the casing string in accordance with the accepted method (Figure 5.20). Each element of technological equipment performs its own functions and is exposed to different loads both during the running of the casing and during its cementing.

5.4.1 Casing Shoe

In order to protect the bottom of the casing from collapse during running (due to cavernousness of the wellbore), a thick-walled pipe, up to 1 m long, is equipped to

5.4 Casing Hardware

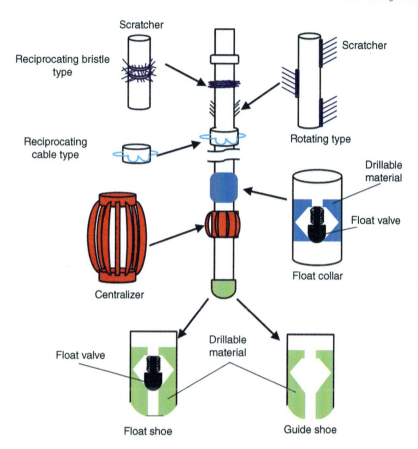

Figure 5.20 Casing equipment.

be called a **casing shoe**. The shoe has different designs and types, the most common of which are listed as follows:

- A **guide shoe** is screwed onto the bottom end of the first casing pipe. Typically, a cone-shaped or spherical plug of easily drillable material such as concrete, wood, cast iron, steel, or aluminum is inserted in the bottom of the guide shoe. It is also possible to make side holes in the shoe ring or plug, through which the slurry is pumped into the annulus (Figure 5.21).
- The **reamer shoe** uses reamer nozzles that allow the wellbore to be slightly enlarged by cutting off protrusions and irregularities (Figure 5.22). In turn, increasing the gap between the borehole wall and the casing increases the likelihood of completing the cementing job effectively.

Figure 5.21 Guide shoe АО "ПО СТРОНГ".

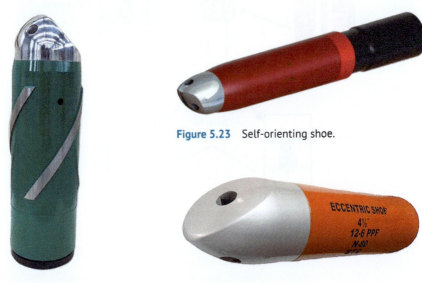

Figure 5.23 Self-orienting shoe.

Figure 5.22 Reamer shoe.

Figure 5.24 Lipstick shoe.

- The **self-orienting shoe** in this design provides free rotation of the eccentric nose section on a high-strength bearing section (Figure 5.23). When lowering a string with this type of shoe, an increase in axial load results in a translational motion with the body rotating, which helps bypass ledges, bumps and cavings, allowing the string to be lowered to the target downhole.
- The **lipstick shoe** with an eccentric head. The design resembles that of a self-orienting shoe, but this type does not rotate the nose part and requires the entire casing to rotate for forward motion (Figure 5.24).

In case there is a danger of plugging the guide shoe (rockfall, rockslide, and bottom hole contamination), a nozzle is installed above the shoe (below the check valve), which is a 1.5 m-long piece of pipe with holes arranged in a screw line for the exit of drilling and cement slurry into the annulus. It is also possible to use a **float collar**, which also serves as a connector between the annulus and the internal space of the well. The float collar allows flow channels for the cement slurry to be opened at the right moment and then closed again. The location of the sleeve is determined in advance by the length of the cementing intervals.

5.4.2 Check Valve

The next element of the casing assembly installed above the casing shoe is one or two check valves. Installation of check valve solves the following tasks:

- It prevents the backflow of slurry into casing during the cementing process.
- Cementing plugs can land on the check valve.
- Automatic filling of running casing string with drilling mud from a well without its overflow from casing to wellhead.

Check valves are subdivided according to their operating principle:

- **Blind valves:** Prevent fluid inflow into casing string during running-in;
- **Differential:** Fill up casing at a certain pressure difference between casing and annular space. In this case, the design excludes the possibility of mud reverse circulation.
- **Butterfly valves:** Provide constant filling of column with solution during running, allowing backflushing of well by reverse circulation.

In terms of construction, the check valves are subdivided into the following groups:

Flapper-type check valve. When pumping liquids due to pressure acting on it, flapper (disc) moves away from seat, and after circulation, termination spring presses it tightly to the seat again (Figure 5.25). Check valves of this type are generally unreliable.
- **Ball-type check valve.** In this design, the separating element is a metal ball (Figure 5.26).
- **Poppet-type check valve.** In this design, the separating element are throttles located below the ball valves, provide filling of the running casing with liquid from the well (Figure 5.27).

It should also be noted that collars and shoes equipped with check valves in English-language literature are singled out into a separate group called float equipment.

A stop collar, which is a cast-iron thick-walled plate with a reduced bore, is installed 20–30 m above the shoe to stop cementing plugs. Sometimes stop collars are made of cement.

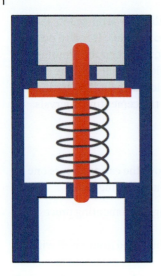

Figure 5.25 Flapper-type check valve.

Figure 5.26 Ball-type check valve.

Figure 5.27 Poppet-type check valve.

5.4.3 Centralizer

Centralizers are installed on casing to maintain alignment of wellbore and running casing and to create favorable conditions for uniform distribution of cement slurry over an annular gap. Centering effect depends on the correct choice of centering interval along wellbore and distance between centralizers on a string. Centralizers are placed at the most critical sections of casing where reliable isolation is very important (pay zone, bottom of casing, etc.) In practice, centralizers are set every 10 m in intervals of pay zones and places of wellbore curvature, in other intervals of cementing centralizers are set every 50 m.

There are the following types of centralizers:

- **Bow-string centralizers.** Standard centralizers for normal well conditions. This type of a centralizer is comprised of two shells connected by a calculated quantity of spring slats of a certain cross-section. Centering is provided by the specified spring properties of the centerline bars. Restriction of longitudinal movement on the pipe body is provided by means of a locking ring and a clamp. They are available in slip-on and hinged configurations:
  - *Slip-on centralizers* are of one-piece construction and require preinstallation on the casing (Figure 5.28a).
  - *Hinged centralizers* consist of two halves hinged together (Figure 5.28b). It is easy to slide it onto the casing above the wellhead when running the casing.
- **Rigid centralizer**

  The strongest type of centralizer, providing forced alignment of the string due to its one-piece rigid design (Figure 5.29). This type of a centralizer, with some modification of design, can combine the functions of turbulators.

  Turbulizing centralizers provide for turbulization of fluid flow in annular space during cementing, thus significantly increasing efficiency of displacement and substitution of drilling fluid with cement slurry in cavernous sections of wellbore, and also eccentric position of drill string in the wellbore. The area of application of turbulators is universal.
- **Roller centralizer**

  This type of centralizers also has one-piece rigid construction with rollers fixed on blades (Figure 5.30). Small contact area of rollers with casing and wellbore wall significantly reduces friction coefficient and sticking of tubing in areas with high differential pressure.

Figure 5.28 Bow-spring centralizers (a) slip-on centralizer; (b) hinged centralizer.

(a)

Slip-on centralizer

(b)

Hinged centralizer

Figure 5.29 Rigid centralizers.

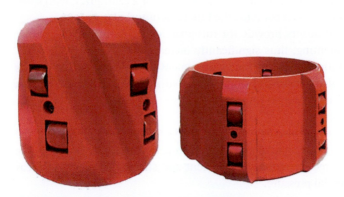

Figure 5.30 Roller centralizers АО "ПО СТРОНГ".

5.4.4 Turbulator and Scratcher

Turbulators are used to create turbulent flow in the borehole for better mud displacement (Figure 5.31). As centering elements also have a turbulizing effect, they are often used instead of a turbulator and equipped with turbulizing blades to enhance the effect. Turbulators are mainly installed in the zones of caverns and productive intervals of the formation. In order to protect turbulizing blades from damage, they may also be installed in a set with centralizer, above it at a distance of 1.0–1.5 m. No more than two turbulators may be installed on one casing.

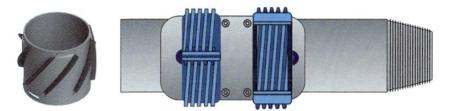

Figure 5.31 Turbulator. www.zavodnpo.ru

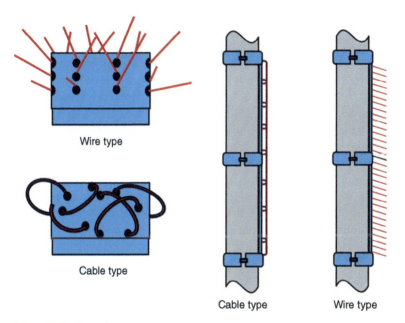

Figure 5.32 Scratchers.

Scrapers are installed on the casing in order to remove the clay crust from the walls of the well (Figure 5.32). Scrapers are subdivided into: wireline, rope, rotating, nonrotating, etc.

5.4.5 Cementing Plugs

Cementing plugs are semirigid barriers between cement slurries and drilling fluids with the following main functions:

- Separation of cement slurry and drilling mud
- Cleaning of wellbore space from mud residue
- Indication of the end of cementing slurry injection into annular space

Figure 5.33 Upper (black) and Lower (red) cementing plugs. MED, Inc.

Today, cement plugs are durable structures made of corrosion-resistant materials, but originally cement plugs were made of burlap, wood, and leather.

Despite the external similarity, the top and bottom plugs differ significantly in their internal design and operating principle (Figure 5.33). Bottom plugs were designed for cement slurry advancement, which requires a pass-through channel in the design. Such a channel is formed when the elastic diaphragm in the core of the bottom cementing plug breaks when it is seated in the stop collar. In the upper cementing plug, the core is steel and is destroyed not by the action of differential pressure, but only as a result of drilling out. In spite of the effectiveness of a lower cementing plug, it still should be used with extreme caution in case of the presence of additives in cementing slurry to prevent loss of circulation. At high concentration of these additives, the check valve may be clogged. Top and bottom cementing plugs, as a rule, are painted in different colors, as in case of wrong sequence of their application the cement will harden in the casing and the only way out will be its drilling out. In the offshore fields, it is possible to use plugs of a slightly modified design. Cementing plugs can be provided with fixing structures, which prevents their rotation, making it easier to drill out.

5.4.6 Cementing Head

A cementing head is a construction for fitting the wellhead of oil and gas wells in order to connect the casing with the injection lines of cementing and drilling fluids, to install cementing plugs and additional cementing equipment.

Cementing heads are classified as follows:

- depending on the type of set string:
  - for casing
  - For drill pipes
- According to the number of plugs:
  - cementing heads with one plug (Figure 5.34a)
  - cementing heads with two plugs (Figure 5.34b)

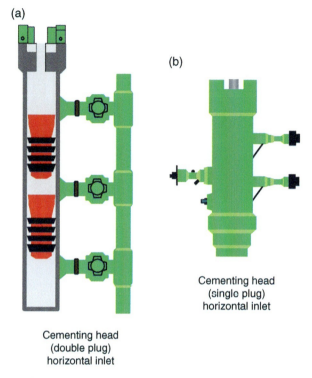

Figure 5.34 Cementing head (a) double plug cementing head; (b) single plug cementing head.

5.4.7 Screening Devices and Cement Baskets

Screening device – a device for limiting sedimentation processes in cement slurry, filling wellbore space behind the casing (Figure 5.35). The Screening Device consists of spring petal cap, interacting with screen element – rubber petal collar. After cementing is complete the petal collar and the shell of the screening unit, contacting tightly enough between themselves and with the walls of the well, create support for sedimentation of cement slurry. On it a compacted cement bridge is formed spontaneously, preventing together with the screening device from lowering a column of cement slurry in the wellbore.

Cementing basket – consists of a flexible steel spring arc welded to a collar installed on the casing or liner along intervals of weakly cemented rocks to prevent absorption of cement slurry under hydrostatic pressure (Figure 5.36). The basket arrangement prevents cement slurry from flowing downward, reducing hydrostatic pressure. In fact, hydrostatic pressure in this case presses not on the formation, but on the cementing basket.

Figure 5.35 Screening device.   **Figure 5.36** Cementing basket.

5.5 Remedial Cementing Equipment

Remedial (squeeze) cementing is cementing, which is carried out within the framework of workover operations in case of detection of leakage, violation of string tightness or inflow of formation water into the well through the perforation holes together with oil or gas.

All remedial cementing methods consist of pumping a calculated amount of cement slurry into the well in order to seal the string or isolate the pay zone.

Remedial cementing tools are mechanical or hydraulic devices, which isolate certain sections of the casing from the pressure caused by hydrostatic pressure of the cement column or injection of cement slurry. These devices are available in both retrievable and drillable variations, and which type of construction to use in a particular well is determined by downhole conditions.

5.5.1 Cased – Hole Remedial Cementing Equipment

5.5.1.1 Packers for Squeeze Cementing Operations in Cased Wells

These packers are mainly used to isolate the upper part of the casing and the wellhead. They are available in both drillable and retrievable designs, which differ significantly in appearance and operating principle.

Retrievable packers can be set and released many times, which makes them practically irreplaceable in operations for testing selective formations or cementing several zones. This type of packer is usually run on a tubing string and is activated in two ways:

- By increasing the pressure acting on the packer (**compression-set**) – working on the principle of deforming the sealing element and isolating individual parts of the borehole below the weight of the tubing string.
- By increasing the tension acting on the packer (**tension-set**) – in this case, the opposite process takes place, the compression load on the packer is reduced. These packers are less common due to the potential difficulties associated with their removal. Tension packers are typically used at shallow depths, with the packer typically descending below the installation depth, thereby providing the necessary lifting distance for activation.

Compression packers are usually more versatile and are therefore preferred if there is sufficient tubing height, as the weight of the tubing should allow for the minimum compression pressure required to activate the packer (40–90 kN). Compression packers are released by simply lifting the tubing.

Retrievable compression packers include many design features for increased productivity and versatility:

- Hydraulic clamping brackets to counteract the force generated by increasing pressure under the packer.
- A bypass (bypass) valve that can be opened or closed in the installed position, allowing circulation without unlocking the packer. In addition, the presence of a bypass zone reduces piston effects and allows back circulation of excess cement slurry without an excessive increase in pressure.

Drillable packers are often used instead of retrievable packers to prevent backflow of cement. They are better suited for situations where there may be communication with top perforations or casing problems that could cause the retrievable packer to cement in the well. A slip collar or flow-through valve is provided to control fluid placement and maintain final squeeze conditions. The sliding collar is controlled by raising and lowering the tubing. This type of packer does not allow fluid flow in either direction.

There are three basic methods of installing expandable packers:

- installation of the packer by means of a wire – allows precise control of the depth. In this case, the packer is lowered to the desired depth and installed by electrically igniting a slow-burning charge in the installation tool, which is retrieved together with the wire after installation of the packer.
- installation of the packer with a drill string or tubing.
- Installing the packer with a coiled tubing.

A separate group of packers is the so-called dummy packers (bridge plugs), which are used to isolate the casing below the zone to be treated. They are also available as a drillable or retrievable packer.

5.5.1.2 Wellbore Tools for Tubing Pressure Testing and Pressure Equalization in the String and Annulus

Tubing testers are down hole valves used to test tubing for leaks. They are commonly used in squeeze cementing operations because of the potential problems caused by leaking connections. High differential pressure in the event of leaking connections will result in localized dehydration of the cement, which, in turn, will increase pressure and create a false impression of the process to the cementing crew. In the worst case, leaky joints may lead to tubing plugging. These devices are usually equipped with a fully opening flapper mechanism that allows the use of wireline tools (e.g. perforators) on the well. Tubing integrity testers are installed above the packer and run in the open position to ensure the well is filled. A ball seat is also possible, of course, but this would limit the diameter of the tubing used and require a reverse circulation to bring the ball to the surface.

Tubing unloaders are bypass valves installed in the tubing string to provide alternate passage of working fluids and are actuated by raising or lowering the tubing.

5.5.2 Open Hole Remedial Cementing Equipment

In open hole wells, inflatable service tools are predominantly used, which are almost analogous to conventional tools and have been used in the oil and gas industry for more than 50 years. Almost the entire range of equipment for secondary cementing in inflatable design is available on the market. Inflatable packers use a liquid-inflated rubber element to activate the packer and seal off a portion of the casing or open hole. They have a smaller outside diameter than their conventional counterparts and increase in size by up to 250% when inflated. Such tools are especially useful in open hole wells of indeterminate size. Like conventional packers, inflatable packers are made in both drillable and retrievable forms, allowing the same operations to be performed as conventional equipment. The advantage of this equipment is that it does not experience compression or tension. The design of the tool allows it to hold significant internal pressure. The tool is run into the well using drill pipe, cable, or coiled tubing.

6

Primary Cementing

6.1 Planning

Each cementing operation has a set of specific objectives, combining technical conditions, project economics, and the legal framework of whatever state the well is located in. The cementing team's engineering staff primarily considers three basic types of data:

- Depth and design of the well – this set of data determines the density, rheology of the cement slurry, and injection rate.
- Reservoir conditions (temperature and pressure) – this data set determines the need for cementing fluid additives.
- Physical and chemical properties of the drilling fluid in the well-this type of data have a multifactorial effect on cement slurry density, displacement behavior, and chemical compatibility of the fluids used.

6.1.1 Depth and Design of the Well

Engineers must have a minimum set of information about the well when planning the primary cementing operation, including:

- True vertical depth of the well
- Measured depth of the well
- Deviation angles and azimuth
- Casing size and weight
- Borehole length and diameter
- Casing type

Depth data are especially important because they strongly influence temperature, fluid volume, and hydrostatic and dynamic pressure. Wells with high inclination angles are particularly challenging and require rigorous control of drilling fluid displacement and mud stability. No well can be exactly vertical; most wells are curved and tilted in different directions. If the exact well path is not known, the casing will not be properly centered and will touch or be tight against the wellbore walls at various locations, greatly reducing mud removal efficiency. Therefore, it is critical to have accurate information about the wellbore trajectory. Drilling with borehole assemblies that contain tools for measurements while drilling (MWD) or logging while drilling (LWD) is becoming increasingly common, and this now ensures that borehole trajectory information is available. However, in cases where such information is not available, engineers often take into account a small deviation angle (typically 3°).

The casing diameter and type, as well as the open hole diameter, are chosen according to expected downhole conditions and completion configuration.

The size of the open hole is dictated by the size of the drill bit. In a real wellbore, an open hole is rarely "calibrated" (i.e. perfectly round and cylindrical). Zones or formations composed of soft and unconsolidated rocks (e.g. shale) are usually unstable and collapses are not uncommon, which, in turn, leads to a high cavernosity coefficient. The cavernosity coefficient plays a critical role, sometimes greatly complicating calculations of the mechanics of drilling fluid displacement and the required volume of cementing fluid. The method used to determine the diameter change along the well path is called cavernometry, which is essentially a special case of logging, where the change in the average actual diameter along the wellbore is measured, which is the diameter of a circle equivalent in area to the cross-sectional area of an irregularly shaped well. The most widespread are caliper tools with down hole lever-type measuring elements (arms) and resistive transducers of linear displacements of arms into an electrical signal. In such devices, the position of measuring arms is mechanically connected with the adjustment of variable resistors; during lowering the device into the borehole, the arms are in the folded and fixed position, and when moving the device up the borehole, the lower ends of the arms rest against the borehole walls, and the upper ones act on the adjustment mechanism of resistors, changing their resistance in proportion to the deviation value of each arm. Recording these values as a function of well depth is presented in the form of two logs, and their half-summary – as a caliper log, which contains all the information about the variations of the well profile depending on the depths.

The calipers are manufactured in different variations, usually differing in the number of arms (from 2 to 6), which should be taken into account when carrying out measurements (Figure 6.1). Some peculiarities of wellbore volume measurements depending on the form of well cross-section and the type of caliper used are

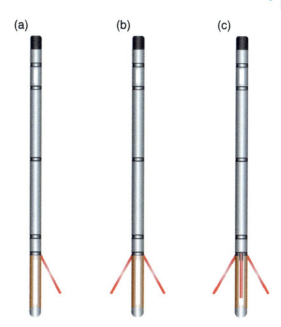

Figure 6.1 Caliper tool types: (a) one-arm caliper; (b) two-arm caliper; (c) three-arm caliper.

shown in Table 6.1. It should be noted that there are methods based on the use of ultrasonic technology, but due to the complexity of the requirements to downhole fluids, they have not found wide application. Another alternative method is the use of tracer fluids, but it also has a significant disadvantage, due to the necessity of the completion of at least one complete circulation cycle. However, it is often common practice to inject a certain amount of excess cement slurry to ensure complete filling of the annulus, but this practice is not without disadvantages, since there is no standardized technique to determine the amount of excess cement slurry to be injected.

With the advent of MWD and LWD systems, the geometric dimensions of an open borehole can be determined from acoustic measurements, radioactive logging, or electrical resistivity measurements in conductive drilling fluids. Combinations of LWD measurements are particularly useful for detecting wellbore stability and trajectory problems. This early diagnostic information allows engineers to suggest ways to correct problems and take immediate action before the problem worsens.

The choice of casing weight, grade, or metal is determined by the mechanical loads the casing will be subjected to, the corrosive nature of the formation fluids, the possible high pressure differentials on the casing wall resulting from large differences in cement slurry and displacement fluid density or from high pressures during final displacement or casing testing.

6 Primary Cementing

Table 6.1 Influence of the type of caliper used and the shape of the wellbore cross-section on the results of wellbore volume measurements.

Caliper type	Circular shape of the wellbore cross-section	Oval cross-sectional shape of the wellbore
Two-arm caliper	The correct calculation	Incorrectly calculated wellbore volume
Three-arm caliper	The correct calculation	Calculated wellbore volume is less than the actual value
Four-arm caliper	The correct calculation	The calculated wellbore volume does not reflect the actual value, but is acceptable for practical application
Six-arm caliper	The correct calculation	The calculated wellbore volume does not represent the actual value, but the resulting data represent the most accurate values that are possible to measure

6.1.2 Reservoir Conditions

Reservoir conditions play a significant role in planning cementing operations. So, it is necessary to have data on reservoir fluids, pressures, and temperatures, as well as the likelihood of water and gas breakthroughs for the proper selection of cementing fluid additives. Reservoir pressure and temperature are especially critical in this case.

6.1.2.1 Pressure

It is essential to have data on fracture pressures along the entire well path, which is accomplished by logging. However, in the absence of such data, engineers often use the density of the mud used to drill the well. This is essentially a very simple and effective method based on empirical data. The logic behind this solution is that if a particular density of drilling fluids does not cause hydraulic fracturing, it is highly unlikely that an identical (or slightly higher) density of cement will. Alternative sources of reservoir pressure data may include reservoir test data, data from various stimulation operations, and LWD measurements during drilling.

6.1.2.2 Temperature

Wellbore temperature is one of the most important factors affecting cementing design and slurry formulation. The engineer needs to have information about the temperature difference between the bottom and top of the cement string, bottom hole circulating temperature (BHCT) and BHST.

- BHCT – theoretically reflects the maximum temperature, to which the cement slurry is exposed during injection, i.e. being in motion (circulation). Down hole the cement slurry is most strongly affected by both temperature and pressure, which affects both time of cement slurry setting and cement strength. However, circulating temperature at bottom hole is always lower than the static one, as circulating cement slurry simply does not have enough time to heat up. Circulating temperature is called bottom hole circulation temperature (BHCT), and it is used in laboratory tests for the determination of setting time. This is typically calculated using a set of temperature charts published by the American National Standards Institute (ANSI), but it should be noted that ISO/API standards do not account for abnormal wellbore conditions and assume that wellhead temperature is constant at 80 °F (26.7 °C). In cases where ISO/API standards cannot be applied, engineers rely on various proprietary methods or the standards of the service company they represent. There are no uniform and generally accepted methods of calculation for such cases. Direct extrapolation of LWD/MWD drilling data is also impossible, as both geometry and fluid flow rates will be significantly different during cementing. Nowadays, computer simulators that simulate the physics of heat transfer under dynamic conditions

are widely used. These methods take into account all well parameters important for heat transfer, including annular space geometry (flow and contact area), fluid rheology, flow rate, and injection temperature, and can predict temperatures at various points in the well, which is especially valuable when cementing deep, deviated, or horizontal wells.
- BHST is an indicator that reflects the temperature to which the cement slurry will be heated at the bottom hole when there is no circulation. This parameter plays a vital role in predicting the rate of cement strength development and its long-term stability. Typically this parameter is calculated from the average geothermal gradient, but logging data can also be used. Although ISO/API standards dictate that cement strength be measured at static downhole temperatures, computer simulators can predict the rate at which cement slurry temperature rises in the well, and for critical jobs, following such graphs may be preferable.

6.1.3 Drilling Mud Parameters

Cement slurries are usually incompatible with drilling fluids. The mixing of these fluids leads to negative changes in the rheology of the fluids and, consequently, to incomplete displacement of the drilling fluid from the annulus. To prevent mixing, drilling and cementing slurries are usually injected with buffer and flushing fluids. Such fluids are designed to be chemically compatible with both drilling and cement slurries. For oil-based drilling fluids, it is almost always necessary to use solvents or surfactants in the flushing/buffer fluids to improve compatibility and remove the oil film from the surface of the casing and formation. Complete displacement of the drilling fluid is one of the main factors ensuring the quality of cementing operations. The efficiency of displacement increases with reducing the viscosity of the drilling fluid. Data on the compatibility of drilling, plugging, and buffer fluids are obtained by laboratory tests according to procedures defined by the International Organization for Standardization (ISO) and the American Petroleum Institute (API).

6.2 Slurry Selection

When selecting the composition of the cement slurry, the following series of factors should be taken into account:

6.2.1 Density

Despite the fact that initially there is an impression that cement slurry density is dictated by reservoir pressure and hydraulic fracturing pressure indicators, in practice, this process has a complex character. For example, in order to reduce costs of cementing jobs, it is a common practice to increase mud yield by using lighter and expanding cements, which certainly has a negative effect on mechanical properties of the cement rock.

In practice, cement slurry density is determined by the following factors:

- Hydraulic fracture pressure
- Formation pressure
- Cement composition
- Economics of the project
- Type of well

6.2.2 Compressive Strength and Mechanical Properties

Modern level of development of cement industry and additives to it allows stating the fact that high mechanical properties of cement rock do not depend on density of cement plugging solution.

Hard cement stone can be produced even at a very low density of slurry by means of regulation of dry cement particles' distribution or the use of certain additives. In fact, the practical importance of cement stone strength has been overshadowed by indicators such as Young's modulus and Poisson's ratio, which were previously ignored in the context of well cementing. These criteria describe both the elasticity of the cement stone and its ability to withstand temperature, pressure, and tectonic fluctuations during the life of the well. It is worth noting that despite the recognition by most experts of the importance of these criteria in the context of well construction, these parameters in the industry standards, regulations, and national laws have not yet been sufficiently studied, and the mechanical properties of cement stone are still described by the compressive strength. The legal framework of almost every state regulates cement stone strength values for shallow wells to provide the necessary isolation of drinking water sources. Cement stone strength values allowable under different state laws vary, but there are generally accepted recommendations for cement stone strength values prior to drilling the next interval (500 psi [3.5 MPa]) and perforating (2000 psi [14 MPa]). In the case of cementing operations in deepwater, low-temperature wells, the rate of strength set is often more important than the ultimate strength of the cement.

6.2.3 Formation Temperature

Reservoir temperature is a key factor in both the drilling fluid and cementing fluid selection process. At reservoir temperatures above 110 °C, the composition of the cement slurry must be modified to maintain sufficient stability of the cement sheath over a long period of time, preferably for the life of the well. Most often quartz flour is added to the solution composition for this purpose. As an alternative, it is possible to use cement systems specially designed for use in high-temperature wells. It should be noted that low formation temperatures may also cause considerable complications during cementing operations. For example, in permafrost areas, cements with low hydration heat are used, thus minimizing the thawing of frozen formations, which ensures the stability of the borehole structure.

6.2.4 Cement Slurry Additives

Additives for regulating cementing properties were discussed in detail earlier, but nevertheless, it is necessary to note the multifactoriality of selecting one or another additive. For example, reducing cement slurry fluid loss is critical when isolating productive formations or in situations where the annular clearance is small. Additives that control this parameter usually increase the viscosity of the cement slurry, and a different dispersant additive may be required to keep the slurry pumpable. Dispersants in turn increase thickening time, and in synergy, with retarders, they can lead to a significant increase of this parameter at best, and in some cases – to an increase of free water content, or even to the instability of the obtained slurry. Other additives may also be incompatible, with sugar-based retarders having a negative effect on fluid loss control additives. It is almost impossible to list all the sets of incompatible additives. In this regard, it is very important to conduct laboratory tests to establish the compatibility of additives used, both in formation and atmospheric conditions.

6.2.5 Cement Slurry Design

This process is complex and the procedures followed may vary from company to company, but the algorithm remains the same. The process begins by reviewing existing experience with this or an identical well, to develop an initial cement slurry composition that is expected to meet the required performance criteria. However, the variability of physical and chemical properties of both cements and cement additives requires actual laboratory tests to verify the predicted results. The laboratory testing process is quite time-consuming, but experienced engineers and laboratory personnel can significantly reduce the number of tests required to obtain the optimum formulation for a given set of down hole conditions. Once the desired formulation has been selected, the laboratory tests are repeated, but on cement and additive samples taken directly from the well, to prevent problems resulting from changes in the physical and chemical properties of the materials or from contamination during transport and storage.

6.3 Theoretical Basis of Mud Displacement

A high mud displacement index is a critical factor for effective cementing operations.

The main objective of primary cementing of wells is to ensure long-lasting and complete isolation of formations behind the casing. To achieve this goal, drilling fluid and buffer/flushing fluid (if used) must be completely displaced from annular space and replaced with cementing fluid, so that consolidated cementing fluid acquires the necessary mechanical properties to provide quality formation isolation during well lifetime. Incomplete displacement of drilling mud may lead to the formation of voids or channels in cement stone, which, in turn, compromises the integrity of the insulation and leads to a number of complications during well operation. The most common complications include casing overflows, poor formation-casing adhesion, various corrosion processes, etc.

For the engineer, the essence of the drilling fluid displacement process is to optimize casing centralization, select the sequence of working fluids, determine the volume and properties of each, and determine the injection rate. These are often the only variables that the engineer can control. The success of the work carried out is most often assessed by means of various logging operations.

From a chronological point of view, mud displacement begins after the completion of drilling operations and is as follows:

1) In order to give the required properties to the drilling fluid, the borehole is flushed out and the drilling fluid is treated before the drill string is extracted – this process is called mud conditioning.
2) Logging is carried out in the borehole. During this process, the drilling fluid is mostly static.
3) The casing is lowered and the mud is conditioned again.
4) The cementing operation is started, the cementing fluid is injected, and the drilling fluid is displaced.

Despite the fact that this issue has been dealt with by researchers almost since the early years of mechanical drilling, it is unfortunately still one of the most complex and challenging issues in the oil industry. This is because experimental and theoretical approaches suffer from serious limitations, some of which are listed here. One of the key parameters in the mud displacement process that is difficult to reproduce is the ratio of casing length to annular clearance. In laboratory experiments, it is almost impossible to exceed values of this parameter greater than 500, while under real conditions, the order of magnitude of this parameter is 10^4. This does not allow to study the influence of gravitation in the eccentric annulus to the full extent. Of course, it is necessary to understand a simple fact, that taking into account all factors, influencing displacement of the drilling fluid, is a very difficult task and practically is not realizable under real conditions. For this reason, it is necessary to be very careful about the extrapolation of experimental results to field conditions. During the last years, the computer simulation methods have been widely used for solving this problem, but it should be noted that all these methods and approaches are based on predictive models and the accuracy of these predictions is still limited.

It should also be kept in mind that to date, there is no direct method of measuring the mud displacement coefficient and indirect measurement methods are used, the results of which are highly dependent on the interpretation of the data obtained.

6.3.1 Preparing the Well for Running Casing

Well cementing is an integral part of the drilling process and this process has a significant impact on the quality of cementing operations. For example, a high coefficient of cavernousness will virtually eliminate the possibility of good casing centralization, which in turn will reduce the mud displacement rate.

An inadequate drilling fluid composition can result in rock washout, formation of a thick filter cake or drilling cuttings sedimentation, which can be difficult or impossible to remove, regardless of the displacement mode. While even good drilling practices do not guarantee successful cementing operations, they can largely prevent many complications by ensuring optimal wellbore characteristics for subsequent operations. Optimal wellbore characteristics include:

- Controllable borehole pressure values
- Smooth borehole walls with only minor cavings
- Correspondence between borehole diameter and bit size
- Stable borehole (i.e. no erosion or cavings)
- The wellbore free of drill cuttings
- Small amount of mud cake on the borehole walls

Unfortunately, this ideal situation is not always attainable and the cementing process must be designed to minimize the impact of poor wellbore preparation for casing running.

6.3.2 Theoretical Basis for Assessing Circulation and Displacement Efficiency

The most common parameter that determines the ability of one fluid to displace another is the displacement efficiency. When the same fluid is used to displace itself, this parameter is called circulation efficiency. Consider an annular space of volume V_{ann} and length L, filled with fluid no. 1 (displaced fluid), flowing

Figure 6.2 Schematic representation of the displacement profile of fluid #1 with fluid #2.

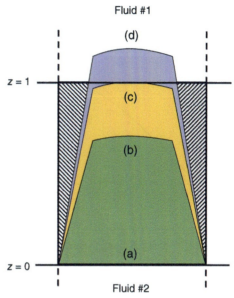

with a given flow rate q (Figure 6.2). At the moment of time $t = 0$, liquid no. 2 (displacing liquid) replaces liquid no. 1 at the inlet of the annular channel ($Z = 0$). At any time $t > 0$, the displacement efficiency $\eta_{displaced}$ is the fraction of the annular volume occupied by fluid #2. In other words, for case (d) in Figure 6.2, the displacement efficiency will be 1 − the shaded area divided by the area between $z = 0$ and $z = 1$. The natural time scale, which allows determining the dimensionless time t^*, is the ratio of the annular volume V_{ann} to the flow rate q.

This dimensionless time is equal to the number of annular volumes pumped. Note that in these definitions, $\eta_{displacement}$ is equal to t^* (case [b]) until fluid #2 breaks through to exit the annulus (case [c]). This time is defined as the breakthrough time – t^*. After the breakthrough (case [d]), $\eta_{displacement}$ approaches but never reaches a constant value less than 1, indicating that the annulus still contains undisplaced fluid no. 1 (Figure 6.3).

a) $t^* = 0$
b) $t^* < t^*_{breakthrough}$ before breakthrough
c) $t^* = t^*_{breakthrough}$ at the breakthrough
d) $t^* > t^*_{breakthrough}$ after breakthrough

Circulation or displacement efficiency is a simple and straightforward concept, but it can sometimes be misleading, especially if the casing is poorly centered. For example, it is possible that an ineffective mud displacement zone occupies most of the length of the borehole circumference, but the area of this zone is small

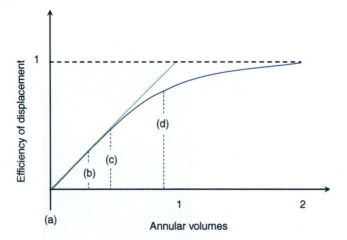

Figure 6.3 Schematic representation of the displacement efficiency curve.

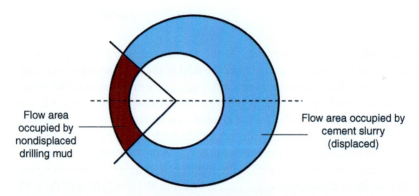

Figure 6.4 Schematic representation of flow area distribution in case of poor casing centralization.

compared to the total flow area (Figure 6.4). However, even such small amounts of unswept mud can create a number of serious operational complications during the life of a well.

6.3.3 Conditioning the Drilling Mud

To ensure drilling efficiency, engineers and laboratory personnel are guided by a number of requirements for the physical and chemical properties of the drilling fluid. In practice, the resulting fluid characteristics often do not always contribute to the efficiency of the displacement process, if not complicate it at all. In fact, such a situation is natural, as the tasks faced by drilling and cementing specialists are fundamentally different. It is for this reason that the mud in the borehole must be conditioned (i.e. changed) prior to the cementing process. Mud conditioning refers, first of all, to the modification of two basic parameters – rheological properties and density of the drilling fluid. The target values of these parameters to which they should be brought are selected individually for each well, depending on its conditions.

Nevertheless, the algorithm used by specialists remains unchanged, the efforts are primarily aimed at reducing the density of drilling fluid (of course, providing the necessary hydrostatic pressure in the well to control its stability), static gel strength, values of yield strength, and plastic viscosity. The aforementioned characteristics are regulated by adding various additives to the mud cleaned of drilling cuttings. It is desirable not to stop the circulation of mud in the well and to pump at least one volume of mud in the well during this period. Ideally, the mud conditioning process should be carried out before the drillstring is removed. Otherwise, there is an increased risk of gelling of the mud due to its being in a nearly static position for too long a period, i.e. during drillstring retrieval, logging and casing running.

It is often the practice to move the drillstring during conditioning to improve mobility of thickened mud, the rotary movement being more effective in horizontal sections of the wellbore, while the reciprocating movement is more effective in inclined sections.

Equal attention should be paid to casing running speed as well because due to the increased flow rate in the annulus, it is possible to fracture the formation. The equivalent flow rate in the annulus (q_{ann}) depending on the casing running speed ($v_{c.r.}$) is determined by the following equation:

$$q_{ann} = v_{c.r.} \times A_{casing}$$

where A_{casing} – cross-sectional area of the casing.

In general, of course, calculating the risk of increased annular pressure is not just a matter of one equation. In practice, this process is not discontinuous and various inertial forces arise that also significantly increase annular pressure not accounted for in this equation. More complex mathematical models are usually used to calculate the casing run rate.

Mud conditioning and circulation should also be carried out after running the casing because during this period long downtime of drilling fluid in a static position is also possible and the risk of formation of the aforementioned negative phenomena is high.

Unfortunately, at this stage, most service companies limit themselves to mud conditioning. One of the most common casings running complications is increased downhole pressure due to an increased concentration of cuttings in the mud, as they are actually scraped by the casing into the wellbore. In order to prevent this negative phenomenon, it is advisable to circulate in the annulus at intermediate depths before the casing reaches the bottom of the well. To date, continuous monitoring of mud circulation remains the best method of verifying the effectiveness of drilling fluid conditioning.

6.3.4 Drilling Mud Displacement

Determining mud displacement efficiency is more difficult than determining mud circulation efficiency, even though it is essentially the same process as mud circulation. The fundamental difference is that the circulation process displaces the drilling fluid, i.e. the fluid is displaced by a nearly identical fluid (note: in conditioning, the properties of the drilling fluid are slightly modified by adding different additives), while the displacement process displaces the drilling fluid by the cement slurry. The velocity profile of the interface between two heterogeneous fluids is very different from that of a homogeneous fluid flowing in the same geometric space at the same flow rate. In addition to the parameters mentioned earlier, mud displacement depends on the physical and chemical properties of the fluids (i.e. density and rheology), their flow regimes, and their possible mixing interaction.

Effective displacement of one fluid by another is the result of the interaction of various forces. Some forces contribute to displacement, while the others hinder it (buoyancy, viscosity, yield stress, static shear stress, etc.). It is also necessary to consider the possible physical and chemical effects at fluid mixing, but it is a multifactor and complicated task. As a rule, many mathematical models of fluid displacement do not consider them. In addition, all these factors are complexly related, for example, buoyancy is a direct function of fluid density difference and well slope, while viscous forces are a function of fluid rheology, wellbore geometry, and flow conditions. In general, the drilling fluid displacement process can be described by two groups of forces: those that promote displacement and those that hinder it; the dominance of one or the other force depends on specific wellbore conditions.

In a wide vertical annulus, where a heavy fluid displaces a lighter fluid at low flow velocity, the buoyancy force will be the main driving force. In contrast, in a sloping and narrow annulus, viscous forces dominate the flow. In both cases, the length of the borehole is also significant.

As mentioned earlier, no laboratory equipment can be scaled to simultaneously reproduce field conditions and account for all independent variables, severely limiting the current understanding of the effects of these forces on mud displacement efficiency.

Of course, research in this area is ongoing and a number of processes and phenomena occurring during mud displacement are now well understood and explained. A few examples are given here:

- All other things being equal and flow rates low, upward displacement of a heavy fluid by a lighter fluid in a vertical channel results in an unstable phenomenon known as a floating plume, which reduces the displacement efficiency. In contrast, when the displacing fluid is heavier than the displaced fluid, buoyancy forces tend to smooth the interface between the two fluids and promote efficient displacement.
- A similar phenomenon is observed when the viscosities of the displacing liquid and the displaced liquid are different. Thus, when both liquids have the same density, but the viscosity of the displaced liquid is lower, formation of viscous fingers is observed, that is, the flow of less viscous liquid penetrates the interface and enters the more viscous liquid. On the contrary, in a similar situation, when the viscosity of the displacing liquid is higher, the displacement efficiency increases.

In general, the application of mathematical modeling techniques remains the most common approach for evaluating the displacement efficiency of a drilling fluid. The models used can be classified according to their complexity:

- One-dimensional models are limited to describing simple steady-state conditions but are nevertheless quite successful for routine operations. The results tend to provide partly quantitative criteria and recommendations.
- Two-dimensional models allow more volumetric, either axial-radial flow (for example, predicting filtration crust thickness along the borehole walls) or axial-azimuthal flow, that is, the formation of unswept mud channels. The data obtained with the use of such models is an excellent tool for predicting drilling fluid displacement efficiency.
- Three-dimensional models are the most advanced and comprehensive tools available, but because of their complexity and high hardware requirements, they remain research tools for the time being. As computational power continues to grow, such tools will eventually find widespread use.

Also of note is the phenomenon known as "free-fall" or U-tube that occurs when injecting dense fluids, such as cement slurry (Figure 6.5). Fluids inside the casing and in the annulus naturally tend to reach hydrostatic pressure equilibrium. The density of cement slurry is usually higher than that of drilling fluid, buffer and washing fluids, which can lead to a hydrostatic pressure imbalance between the inside of the casing and the annulus when it is injected into the well. In this case, there is a "U-tube effect" of the mud in the casing, which creates a vacuum in the upper part of the casing. It is not uncommon for the rate of cement

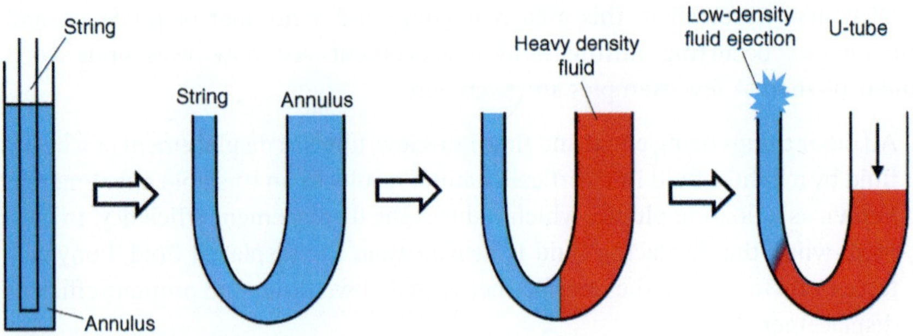

Figure 6.5 U-tube effect.

slurry flow into the casing to be insufficient to keep it full at the initial stage of operation, and the rate of fluid outflow from the well may be much higher than the rate of its inflow. As the casing approaches the hydrostatic pressure equilibrium point, the outgoing flow rate becomes lower than the incoming flow rate, and the casing gradually fills up. In this case, an interesting effect can occur: at some point, the outgoing flow velocity can reach zero and the fluid column in the annulus will be stationary, which can easily be mistakenly interpreted as a partial or complete loss of circulation. However, once the casing is filled with fluid again, the velocities of incoming and outgoing flows will be equal, but even this equilibrium may change before the displacement process is complete.

Thus, in case of using low-density flushing fluid, the annular pressure may decrease as it passes through the casing shoe, which, in turn, will cause another U-tube effect, accompanied by another increase in the outgoing flow rate.

Given the importance of casing fluid velocities and pressures for safe and successful cementing operations, the "U-tube effect" should be considered in any cementing design.

In general, the issue of displacement of drilling mud during cementing operation, as it was mentioned earlier, is rather complicated and complex. To describe all possible models and approaches used in practice today would significantly complicate the reader's perception of the material and for a more detailed study of this process, the reader is recommended to read the list of references given at the end of the book.

6.4 Methods of Well Cementing

Most primary cementing operations are performed by pumping cement slurry through the casing to the bottom of the well, from where it rises up through the annulus to the wellhead. This method of cementing is considered traditional, but due to a variety of problems encountered when injecting wells, this method of cementing is not always applicable. For instance, if the lower interval of the well has weakly cemented rocks, the method of reverse grouting is applied, when cement slurry is injected through the annulus to the bottom of the well and then rises through the casing to the wellhead. It should be noted that this method has disadvantages because of which it has not been widely used. The main and most significant disadvantage of this method is the difficulty of placing the cement slurry in the planned interval and the need for special underground equipment for controlling the upward flow of cement slurry in the casing. For casing strings of bigger diameters, the injection of cement slurry through drill pipes is also widely used, injection of cement slurry into annular space with one or several small-diameter pipes is also possible. Selection of one or another method of cementing depends primarily on specific borehole conditions (i.e. depth, borehole inclination, geological conditions, etc.) Some of the most common methods of well cementing are discussed here.

6.4.1 Cementing Through Drill Pipes

As discussed earlier, working through drill pipe can prevent many of the problems associated with cementing large-diameter casing. In this method, the casing is run into the well with a stab-in float shoe. The casing is placed on the slips and the drill string, equipped with a stab-in stinger (Figure 6.6), is lowered 1 m above the float shoe. Next, the circulation of the drilling fluid in the annulus between the drill string and casing is established. The circulation is stopped and the drill pipe is lowered until the stinger is sealed on the float shoe. Next, the fluid level in the annulus

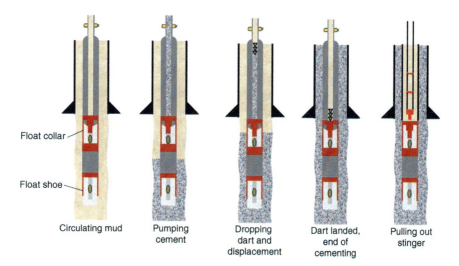

Figure 6.6 Through drill-pipe stab-in cementing.

between the casing and the drill string is monitored. The fluid level should be constant, but minor overflows are allowed. The next step is to pump the cement slurry through the drill string up through the annulus to the wellhead. As soon as the content of drilling fluid in the slurry coming out at the wellhead is insignificant, the process of pumping the cement slurry is stopped. If circulation is lost during cement injection, the process is stopped, and the pressure drop between the outer annulus and the annulus between the drill pipe and the casing must be strictly controlled to avoid casing collapse due to excessive pressure. To seal the annular space between the drillpipe and the casing, a wellhead preventer can be used to maintain the necessary pressure to compensate for the effect of excessive annular pressure. Another possible solution is to inject drilling mud of the required density into the annular space of casing before the operation. Cementing through drill pipe has several advantages: no exact well volume data are required since the cement slurry is agitated and pumped until it reaches the wellhead, virtually eliminating the possibility of overspending. This method also eliminates the need for large-diameter squeeze or cement heads, as well as larger-diameter cementing plugs for casing cleaning. Not unimportant is also the fact that this method of cementing minimizes the contamination of cement slurry with drilling mud.

6.4.2 Cementing Through Small Diameter (Macaroni) Tubing

This method, in fact, is a modification of the reverse cementing method and is most often used in case of loss of circulation when securing large-diameter casing (Figure 6.7).

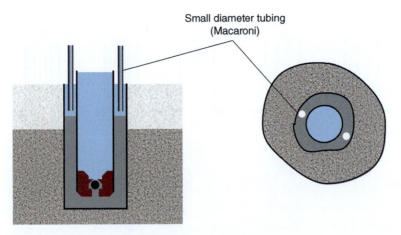

Figure 6.7 Cementing through small diameter tubing.

The tubing string of a small diameter (about 5 cm) is lowered into an annular space between casing and open borehole. Then the tubing string is connected to the cementing unit and the circulation of drilling fluid or water is established. It is necessary to pay special attention to the fact, that the hydraulic losses will be rather high due to the small diameter of the tubing used. In all other respects, the method is almost identical to the traditional method of reverse cementing, that is, the cement is pumped into the annular space until it comes to the surface. It should also be noted that the application of small-diameter tubing – macaroni tubing – is somewhat complicated in offshore conditions.

6.4.3 Single-Stage Cementing

With the development of new ultra-low-density cement systems, the need for multistage cementing has been greatly reduced. The procedure for single-stage cementing is as follows.

After the casing is set, the drilling fluid is circulated for the necessary time to remove the drilling cuttings generated during the semistatic period of drillstring retrieval, logging and casing running. The circulation of drilling fluid is usually done through the cementing head to avoid a prolonged stoppage of circulation after mud conditioning. Under static conditions, the gelling of the drilling fluid can progress quite rapidly, which would significantly reduce its displacement efficiency. In the case of mud gelling, circulation is carried out until a stable density of mud in the well is achieved. Cementing is carried out only after the aforementioned parameters are static.

The bottom cementing plug is one of the key elements of cementing operation. Its functions are to prevent mixing of working fluids and to clean inside wall of casing. Under normal conditions, the cement slurry is virtually immiscible with the buffer slurry due to the identical density of both fluids. However, when using a bottom cementing plug as a separator between the buffer and cement slurry, although the buffer fluid is virtually immiscible with the drilling fluid, the drilling fluid residue that the bottom cementing plug will scrape from the inside walls of the well may significantly contaminate it. In addition, once the cement plug is seated on the stopper ring and the rubber membrane of the plug is ruptured, the cement slurry rushes into the annulus, where it eventually mixes with the contaminated buffer fluid, which can lead to a variety of complications.

To avoid this situation, some excess cement or buffer fluid is often injected. A good solution is to use two bottom cementing plugs separating all three operating fluids from each other, but this method has not been widely used.

In practice, the most widespread schemes of single-stage cementing process with the application of one bottom cementing plug are:

- Lower cementing plug release → Buffer fluid pumping → Cement slurry pumping
- Injection of washing fluid → Release of lower cementing plug → Buffer fluid injection → Cement slurry injection
- Flushing fluid injection → Release of the lower cementing plug → Pumping in the cement slurry

6.5 Multistage Cementing

As mentioned earlier, the method of multistage cementing is not often used today, but nevertheless, in a number of cases the use of this method is the best possible solution:

- When there is a high probability of rock fracturing due to high hydrostatic pressure of cement slurry when cementing a sufficiently long interval of the well.
- In case of higher fracturing pressure of rocks comprising the upper cementing interval.
- In selective cementing of some intervals of the well, while some intervals of the well were not cemented. To date, this approach has not been used.

The reasons for using this method of cementing may also be due to technical capabilities of the cementing team. For example, when a cementing unit develops less pressure than the pressure at the cementing head, or when there is no excessively large fleet of cementing units and mixing machines required for cementing jobs.

Of course, the aforementioned list of reasons for multistage cementing process is not complete, it only reflects the most common ones.

As a rule, there are three standard multistage cementing methods:

- Standard two-stage cementing, in which the cementing of each stage is a separate and distinct operation
- Continuous two-stage cementing, in which both stages are cemented simultaneously
- Three-stage cementing, where each stage is cemented as a separate and distinct operation

Let us consider each of the aforementioned methods in more detail.

6.5.1 Standard Two-Stage Cementing

In addition to the conventional casing assembly (e.g. guide shoes and float collars), a stage cementing collar is lowered to the desired depth. There are several

Figure 6.8 Stage cementing plugs ООО"АЛЬКОР".

standard designs of cementing collars available on the market, but regardless of the design, extreme care must be given when handling this equipment. Careless handling prior to or during installation can result in "egg" or displacement of moving parts, resulting in malfunctioning.

The stage cementing plug (SCP) is a set of equipment consisting of a steel thick-walled casing with aluminum filling and squeeze plugs (first stage squeeze plug, locking squeeze plug), and a free fall opening device (bomb plug) (Figure 6.8). It is attached to the casing assembly by means of sockets or nipple threads. The first stage selling plug consists of several rubber elements of different shapes fastened to each other by a rod-shaped core. The locking sales plug is one piece and consists of a plastic core with a rubberized outer part, the design includes a plastic element for seating on the socket seat. The free fall opening device is also a one-piece aluminum construction with a lead tip for weighting.

Cementing of the first stage is similar to the single-stage cementing method, the only difference being that in most operations, the bottom cementing plug is not used (Figure 6.9).

After a calculated volume of cementing slurry is prepared and pumped, the first SCP is released and sits on the stop collar. It is necessary to have accurate data on the volume of the annular space to determine the correct height of cementing solution rises in the annular space.

After completion of the first stage, the free fall opening device is released, which by gravity sits on the lower seat, then the pressure increases until the fixing pins shear off, making the lower sleeve move downwards and opening the circulation

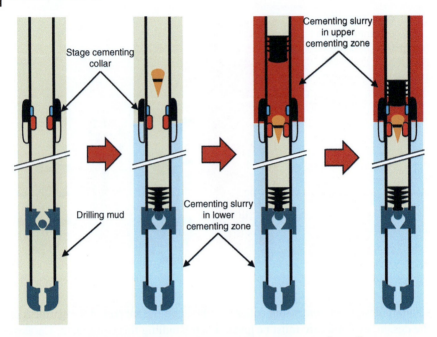

Figure 6.9 Well cementing process with the use of a stage cementing collar.

holes of the sleeve. A sharp drop in pressure at the wellhead is a signal that the operation was successful. It is desirable to reduce as much as possible the time difference between cementing of the first and second stages. It should be noted that the earlier described method uses a mechanical SCP, but there is also a design with a hydraulic mechanism for opening the circulation holes. In such a design, the fixing pins are sheared off by an increase in intracasing pressure. In case of the application of a hydraulic tool, it is necessary to provide circulation of drilling fluid after the opening of circulating holes until its physical and chemical parameters are normalized for cementing operation of the second stage and setting of cement of the first stage. Otherwise, there is a possibility of hydrofracturing of rocks in the lower interval of cementing due to an increase of hydrostatic pressure in the annulus because of additional hydrostatic pressure, created by cement column of the second stage. After the required volume of cementing slurry is pumped, the locking plug is released, and it settles on a special seat in the stage collar, further the displacement pressure is increased to close circulating holes. As a rule for cementing of the second stage cement slurries of low density are used and there are no special requirements for the cement slurry, except for cases when there are high-pressure zones or aquifers in the interval.

6.5.2 Continuous Two-Stage Cementing

This method is presented more as a historical reference since it is practically not used today. The main disadvantages of this method are strict control of working fluid volumes and high hydrostatic pressure values. As a matter of fact, this method does not solve the task of multistage cementing.

Initially, after the cement needed for the first stage cementing is pumped, the cement plug is released, separating it from the displacing fluid (i.e. water or drilling fluid). The volume of displacing fluid is calculated in such a way as to ensure the required level of cement column lift in the annular space (as a rule, to the level of SCC). After the displacing fluid is pumped, the free fall opening device is released, opening the circulation openings of the SCC. Immediately after that the volume of cementing slurry necessary for cementing the second stage is pumped and the closing plug is released, it sits on the socket seat under the effect of flow of the displacing fluid. After shrinkage of the locking plug, the pressure begins to increase and, when it reaches a certain value, causes the plug to move downward and close the circulation holes (Figure 6.10).

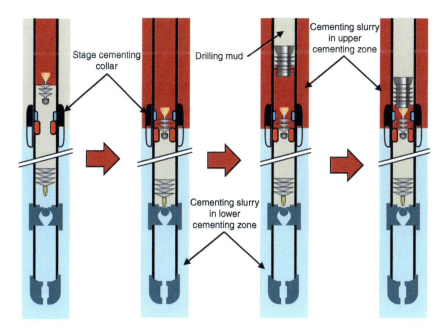

Figure 6.10 Continuous multistage cementing.

6.5.3 Three-Stage Cementing

The three-stage cementing method is also presented as a historical reference since it is rarely used. It increases operating time, complicates the operation and greatly increases the risk of complications. However, at greater well depths and the presence of gas flows, combined with potential casing corrosion problems, this method can be an effective solution. Technically, the process is almost similar to two-stage cementing, but in this case, three well intervals are cemented instead of two. To cement the upper two intervals the cementing plugs of staged cementing are used (Figure 6.11). And cementing of the second and third intervals may be done practically any time after cementing of the first interval is finished.

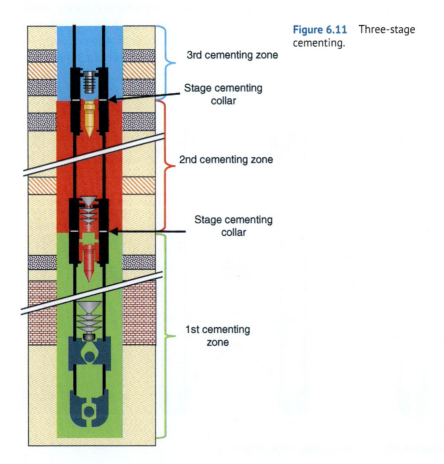

Figure 6.11 Three-stage cementing.

6.6 Liner Cementing

A liner is a string of standard casing that is suspended inside the previous casing string and does not reach the wellhead. The height of liner suspension in the previous string depends on its purpose and varies from 15 to 150 m. A schematic representation of liner classification is shown in Figure 6.12.

- The production liner is suspended in the last casing, lowering it down to the bottomhole. As this liner, in fact, replaces the production casing, the quality of plugging operations is critical to ensure reliable isolation during the whole period of well operation.
- Drilling or intermediate (technical) liners are installed mainly to seal and isolate zones of abnormal formation pressures and circulation losses while drilling. The cementing of this type of liner is challenging.
- A Tie-back Liner is a section of casing installed in the interval from the wellhead to the top of the liner in the well. Reasons for installation of this type of liner are usually: insufficient strength characteristics of the borehole structure or violation of intermediate casing integrity.

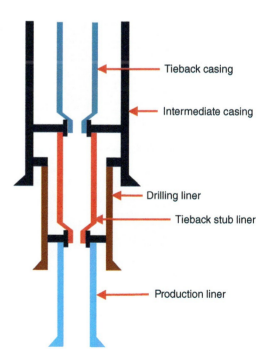

Figure 6.12 Types of liners.

The procedure for running the liner is identical to the procedure for running a standard casing. However, if liner rotation is planned during cementing, the threaded connections are required to withstand higher torques. Use of centralizers is also a practical necessity for cementing liners due to small annular clearances, which increase the risk of jamming while running into the well and reduce efficiency of displacing the drilling fluid. The liner assembly includes a float equipment and, in some cases, a landing collar to provide a seat for the bottom cementing plug. The liner hanger is placed on top of the liner, not infrequently in combination with a packer. The hanger is kept under tension as it descends to the desired depth, which prevents it from buckling under its own weight. The hanger is mounted on the last tube of the liner and is positioned inside the casing at the desired depth. The hanger design provides protection from premature triggering during cementing.

Hangers are classified into three groups according to the setting mechanism:

- Hydraulic;
- Mechanical;
- Hydromechanical (combined).

Line hangers are also classified by design:

- **Rotating.** There is a bearing unit in a design, which provides rotation of a liner in the process of filling slurry, which considerably improves cementing quality.
- **Protected.** This design has elements placed inside special cavities, which fully exclude any damage caused by intensive rotation or rocking of liner during running.
- **Uncemented.** This design allows to fasten the liner without cementing solutions, which reduces the operational costs.

It must be noted that above mentioned are only the most common type constructions of hangers. It is possible to use designs, which imply high tightness of hanger elements or availability of filter isolation unit.

As a rule, the liner is run into the borehole by means of a drill string and a retrievable setting tool (Figure 6.13). This tool performs the following functions:

- Provides a tight seal between the drill string and the liner.
- Holds liner while running into the hole
- Attaches special constructions, which ensure application of cementing plugs.

As it has been mentioned before the displacement ratio of the drilling fluid plays an important role in a successful cementing job, and the liner cementing operation is not an exception in this respect. It is not a mistake to call this type of work one of the most difficult operations the cementing team has to deal with. Small annular clearances, even with a perfectly aligned well, leave almost no room for mistakes. The situation can be exacerbated by even a slight incline in the borehole, which may preclude the use of standard centralizers.

A good way to overcome these difficulties is to use a reciprocating or rotating liner while cementing. However, the following risks of this approach should also be considered:

- There may be complications with disconnecting the drill string from the liner after cementing.
- High requirements are imposed on the strength of the drill string due to the potential danger of its failure during the liner movement.
- Significant increase in resistance when running the liner into the borehole caused by the presence of centralizers.
- High probability of a swabbing effect with formation damage.
- Deformation of the wellbore is possible

Despite the possibility of occurrence of complications described above, this method remains one of the most effective for improving liner cementing quality.

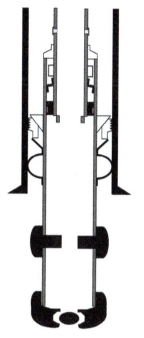

Figure 6.13 The tool for installing the liner and the hanger assembly.

No less important factor in increasing liner cementing efficiency is the use of spacer and washer fluids with maximum possible increase of contact time for both fluids. As a rule, contact time is increased by increasing the volume of working fluids, but in liner cementing conditions, this approach is not always acceptable due to the possible risk of increased hydrostatic pressure. Fortunately, creation of turbulent displacement mode in small annulus spaces is easily achievable even at low pump flow rates without risk of rock hydraulic fracturing, which is actively used in practice.

During liner cementing the cementing head and manifold are mounted on the drill string (Figure 6.14). Once the cementing lines are installed and pressure-tested, a spacer or washer fluid is injected (Figure 6.15). The most widespread is the cementing method using one separating plug released before cementing, although there is a method using two separating plugs (Figure 6.16). The second method is preferable.

After pumping the mud into the drillstring, the lower pumpdown plug is released, which goes through the liner tool and sits on the lower liner-wiper plug. The pressure in the drillstring begins to rise, and when it reaches 8.4 MPa, the pins

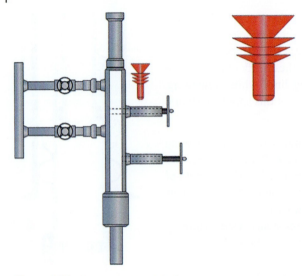

Figure 6.14 Cementing head for liner cementing.

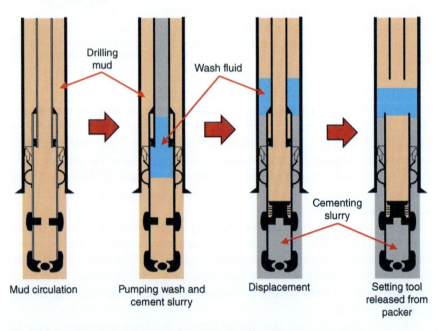

Figure 6.15 Schematic representation of the liner cementing process.

6.6 Liner Cementing

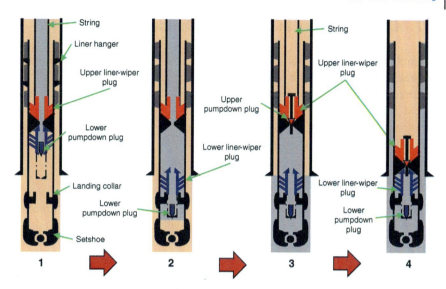

Figure 6.16 Schematic representation of the liner cementing method using two liner plugs.

holding the plug are sheared off. This is essentially the point at which displacement begins, because from that point onward, there is no more cement in the drillstring.

Once released, the liner-wiper plug, as a solid structure, rushes to the landing collar or shoe, depending on the liner arrangement. Upon reaching the landing collar, the pressure begins to rise again, signaling the completion of the operation. To reduce contamination of the cement slurry inside the liner, especially if the liner length is significant, it is desirable to use a system of two separating plugs.

Previously, in practice, when cementing large well intervals using liners, the operation was often performed in two stages. During the first stage, the cement slurry was displaced through the casing shoe, the main condition being that the cement slurry column was raised to the level, at which the previous casing shoe was located, but not overlapping it.

After the first stage was completed, the drill string was removed from the borehole and the cement was allowed to harden. At the end of this time, the drillpipe assembly with the packer was lowered into the hole at a depth above the casing suspension. The second stage was pumping under pressure the cement slurry into the annular space behind the liner. Today, with the advent of high-performance cement systems of ultra-low density and foamed cements, this method is rarely used.

6.7 Critical Factors in Cementing Operations

A number of critical factors must be considered during the planning phase of a cementing operation if it is to be successful. The initial prerequisite is to have accurate data on downhole conditions, and it is equally important to monitor and record gauge readings for further comparison of actual results with design results. The following is a brief discussion of the main factors that must be taken into account during cementing operations.

6.7.1 Volume of Cement Slurry

Because of difficulty to measure exact volumes of annulus, as a rule, after receiving calculated data, some corrections are made to volume of cement slurry, taking into account experience of already performed cementing works in the given field. In case of absence of such information not seldom the calculated volume of plugging solution is increased by 50% or even 100%. In practice there are possible cases of increasing the volume of cement slurry up to 300%, that, as a rule, takes place at deep-water fields. As a whole, practically at each cementing operation there are taken corrections for excessive volume of plugging solution. In some countries, this issue is regulated by legislation.

6.7.2 Displacement of Cement Slurry

To ensure complete cement column in required intervals of wellbore, it is necessary to displace all volume of cementing slurry into annulus. Otherwise, due to insufficient volume, the cement column will simply not reach the upper cementing interval, and the missing volume of cement slurry will remain in the casing. As a result, the annulus will be poorly isolated, and the cement cured in the casing will have to be drilled out. In fact, the calculated values of cement slurry in the annulus may differ significantly from the actual values, due to two most common reasons: lack of accurate data on the casing inner volume and failure to take into account compressibility of the displacing fluid.

The volume of displacing fluid is usually calculated according to the nominal inner volume of the casing. However, the problem is that the inside diameter of the casing has slight differences depending on the manufacturer. Exceeding this value by 1% is enough to cause a critical error in the calculation. In this scenario, an insufficient amount of displacing fluid is pumped and part of the cement slurry remains in the casing without entering the annulus. In order to prevent such costly errors, some companies systematically measure either all pipes in the casing assembly or a statistically significant number of them.

A significant error in calculations is also possible if the compressibility of the displacing fluid is not taken into account. In practice, it means that the fluid will occupy less volume due to its compressibility and part of the cement slurry will also not get into the annulus, remaining in the casing. This parameter is either provided by the drilling fluid manufacturer or determined in the laboratory.

6.7.3 Well Temperature

Bottomhole temperature determines most of the requirements for cement slurry (i.e. composition of cement slurry composition, setting time, cement strength setting rate, etc.) As mentioned earlier, there are two values of bottomhole temperature – circulating (BHCT) and static (BHST).

In terms of dynamics, cement slurry is initially in a dynamic state (i.e. during mixing and pumping of cement slurry) and then in static state (i.e. during hardening) during cementing operations.

Circulation temperature values are used when determining physical and chemical properties of cement slurry in its dynamic state (i.e. thickening time, viscosity, etc.). On the contrary, when describing processes in static state of cement slurry, values of static temperature on the bottom hole (i.e. mechanical properties) are used.

Nowadays data on these parameters are obtained by computer simulation. It is also important to consider temperature of dry cement and mixing water, as hydration reaction also leads to release of some heat. The following is a nomogram which has been used for many years to estimate the temperature of the cement mortar during mixing (Figure 6.17). Today, this nomogram, or a modification, is included in almost every cement mortar design software product.

In general, laboratory experiments should reflect actual temperatures at the surface of the well, since in hot climates they will be very different from the room temperature in the laboratory, meaning that even the cementing slurry preparation process should reflect in-situ conditions as accurately as possible. The concentrations of many of the cement additives (especially the setting accelerators or retarders) will be highly dependent on these conditions.

6.7.4 Well Pressure

Accurate bottomhole pressure readings are also one of the most critical factors for successful cementing operations. Wellbore stability directly depends on hydrostatic pressure created by cement slurry. At the same time, it is necessary to take into account such factors as equivalent circulating density and rheological properties of cement slurry.

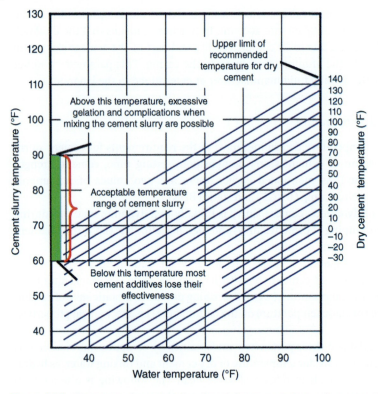

Figure 6.17 Nomogram for estimating the mixing temperature using cement slurry with a water-cement ratio of 0.46 without any additives as an example.

Equivalent circulating density of cement slurry is the density value of non-circulating cement slurry equivalent to the pressure value, which creates hydrostatic pressure of circulating cement slurry combined with differential pressure in annular space. Thus, in fact, cement slurry of any density at high differential pressure (i.e. depending on injection rate) can cause fracturing of the formation.

An increase in cementing solution viscosity in tandem with a high displacement rate will also result in formation hydraulic fracturing due to high hydraulic losses. Therefore, before any cementing operations, a graph is built, showing the maximum and minimum permissible values of hydrostatic pressure (Figure 6.18).

6.7 Critical Factors in Cementing Operations | 141

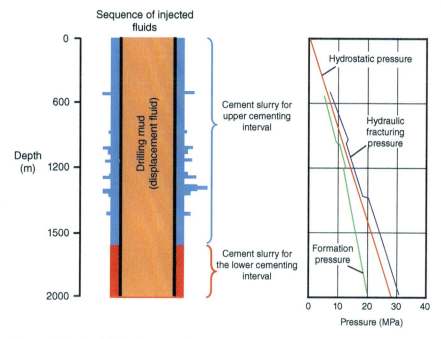

Figure 6.18 Graph of the hydrostatic pressure range.

7

Remedial Cementing

Remedial cementing is a general name for a number of operations that use cementing fluids to control complications that occur after primary cementing. Such complications arise at almost any time during the life of a well, from its construction to abandonment. Let us look at some examples of complications for which remedial cementing is used at different stages of well construction:

- **During the well-drilling stage**, remedial cementing is used mainly to eliminate wellbore integrity problems or to correct defects in primary cementing operations.
- **During well operation**, remedial cementing is used to control production rates (e.g. gas cut, water cut, etc.), repair corroded pipes, or improve zonal isolation in the near-wellbore zone.
- **At the well-abandonment stage**. In this case, the process often presupposes cement plug placement (first of two remedial cementing technologies), but while preparing for the process, it is often necessary to eliminate behind-the-casing flows, which is usually done with squeeze cementing (second of two remedial cementing technologies).

In fact, each remedial cementing operation is a unique engineering task and there is no unified technology that solves the whole range of possible issues faced by the oilfield engineer. Ultimately, this has led to the development of a fairly wide range of remedial cementing methods and techniques that combine the use of various cementing compounds, fluid injection techniques, and cementing equipment.

In spite of a variety of techniques and approaches applied, the remedial cementation comes down to two main technological operations:

- Plug cementing
- Squeeze cementing

Oil and Gas Well Cementing for Engineers, First Edition. Baghir A. Suleimanov, Elchin F. Veliyev, and Azizagha A. Aliyev.
© 2023 John Wiley & Sons Ltd. Published 2023 by John Wiley & Sons Ltd.

Table 7.1 Goals and objectives of remedial cementing.

Plug cementing	Squeeze cementing
Isolation of loss circulation zones	Isolation of loss circulation zones
Sidetracking	Correcting defects in primary cementing work
Creating a base for directional drilling	Isolation of water flow into the well
Isolation of depleted zones of the formation	Reducing the water- or gas-oil factor by isolating intervals of water and gas occurrences
Protection of low reservoir pressure zones during well workovers	Eliminating casing leaks
Providing anchoring for testing in an open hole	Elimination of depleted and unproductive formation zones
Sealing of wells for abandonment	Redirecting the flows of injected fluids

Plug cementing: It is an operation aimed at the creation of a temporary or permanent nonpermeable barrier in a wellbore (whether open or cased).

Squeeze cementing: Basically, it is the process of forcing a cement slurry under pressure much higher than hydrostatic pressure through holes, cracks, or fractures in casing/borehole annulus to provide tightness between the casing and the formation. The cement slurry is pumped until the pressure in the wellbore reaches a predetermined level. This process is largely driven by the formation of a filtration crust as the cement squeezes through, which allows the wellbore to withstand the high pressures encountered during this operation. Once cured, the filter cake forms an almost impenetrable solid mass.

The main goals and objectives of remedial cementing technology are summarized in Table 7.1.

7.1 Plug Cementing

7.1.1 Plug Cementing Techniques

There are several basic ways to install a cement bridge:

- The balance method
- The method using a cement dump bailer
- The two-plug method
- The hydraulic (expanding) packer method
- The umbrella-shaped membrane
- The method with coiled tubing

7.1.1.1 The Balance Method

As a result of its simplicity and lack of requirements for the use of special technical equipment, the balance method is the most common method of installing a cement plug. This method is based on pumping of the necessary amount of cement slurry, buffer, and flushing fluid through drill-string or tubing until a single elevation of fluid rises inside wellbore as well as in annulus (Figure 7.1). The drill string or tubing is lowered into the well to the required depth. In order to avoid drilling fluid contamination, before and after the cement slurry, the rims of spacer and washer fluids are pumped in. After all operating fluids are balanced, the tubing/drilling string is slowly retrieved to a depth above the cement bridge installation, and excess cement is removed. In some cases, fluids with high gel strength are injected to ensure that the cement bridge is properly positioned and to prevent it from sinking down the wellbore. As a rule, such thixotropic fluids as bentonite slurries, cross-linked polymer systems, etc., are used for this purpose. It should be noted that for this purpose, it is also possible to use mechanical devices; for example, hydraulic (expanding packer), umbrella-shaped membranes, etc. Not seldom these methods are classified as separate methods, but, in fact, they are only a variation of the balance method.

7.1.1.2 Cement Plug Installation Using a Dump Bailer

A dump bailer is essentially a container holding a certain amount of cement slurry, lowered into the well on a cable, it is triggered when the cement plug is located below the cementing interval (Figure 7.2). This method allows to control the depth of cementing plug and is relatively inexpensive. The main disadvantage is that the amount of cementing slurry is limited by the volume of dump bailer. However, slurry can be lowered and raised many times. No less important peculiarity of this method of cement plug installation is the prolonged presence of

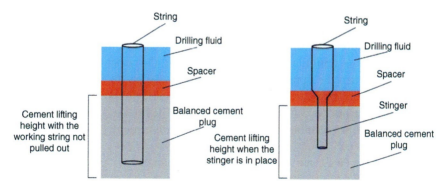

Figure 7.1 Balance method of cement plug placement.

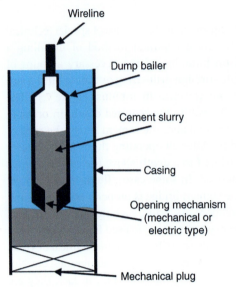

Figure 7.2 Cement plug installation using a dump-bailer.

cement slurry in a static state, which should be taken into account when developing cement composition.

7.1.1.3 Cement Plug Installation Using the Two Plugs Method

This method uses a special set of equipment to allow the cement plug to be installed at the calculated depth with minimal drilling fluid contamination.

In this method, the bottom cement plug is discharged before the cement slurry, separating it from the spacer fluid, and the top cement plug after the cement slurry, separating it from the displacing fluid (Figure 7.3). Both top and bottom cementing plugs are equipped with rubber membranes, resistant to certain pressure, above which they rupture, allowing for cement slurry to pass through them. The drill string/tubing is equipped with a clamping device (locator) located at a small distance from the stinger. When the lower cementing plug is clamped, the pressure increases until the diaphragm ruptures and the cement flows down the string. Next, when the upper cementing plug is seated on the lower cementing plug, the pressure increases again, followed by a diaphragm rupture, and the displacing fluid pushes the cement slurry downward. The final step of the operation is to lift the drill string/tubing and flush out the excess cement.

7.1.1.4 Cement Plug Installation with the Use of Coiled Tubing

The use of coiled tubing for remedial cementing began back in the early 1980s. Since then, the method has gained considerable popularity. Coiled tubing has

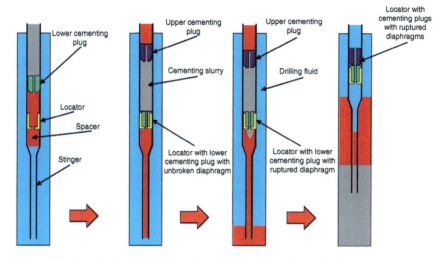

Figure 7.3 Installation of a cement plug using the two-plug method.

proven to be a very economical method of accurately placing the small volumes of cement slurry required to install a cement plug, allowing up to 85% reduction in production costs. Coiled tubing is also widely used in squeeze cementing operations.

7.1.2 Plug Cementing Equipment

The main problem when installing cement plugs is mud contamination of the cement slurry. The volumes of cement slurry used are significantly smaller than in primary cementing operations, and even a small amount of cement slurry contamination, by primary cementing standards, can lead to a dramatic change in setting time and deterioration of mechanical properties. The causes of drilling fluid contamination range from failure to follow mud removal guidelines to high rates of drill string/tubing retrieval once the operation is complete. The primary purpose of cement-plug installation equipment is to prevent these negative scenarios from occurring.

7.1.2.1 Bridge Plug

The bridge plug is placed below the cement plug installation interval and serves as a base of sorts to prevent heavy cement slurry from falling through the less dense fluid beneath it. In the case of a retrievable bridge plug, a common practice is to fill one or two bags of sand over the plug before pumping the cement slurry in order to prevent cementing of the release mechanism. The second most common type of bridge plug is the drillable version. The design of such plugs is similar to that of packers. It is possible to run into the well both with the help of a wireline and by means of drill pipe string. Regardless of their design, bridge plugs do not allow the downward flow of upstream fluids to pass through.

7.1.2.2 Tailpipe or Stinger

To reduce drilling fluid contamination when retrieving the drill string after the operation, smaller diameter pipes, called stingers or tailpipe, are typically used at the bottom of the assembly. It should be noted that the material from which the stinger is made can vary from aluminum to fiberglass. It is not uncommon for the stinger not to be removed after the operation, leaving it in the cement mortar as a reinforcing element. In this case, after the cement hardens, the top of the pipe is destroyed by increasing pressure or blasting.

7.1.2.3 Diverter

In order to avoid the mixing of the cement slurry coming out of the pipe with pre-injected fluids that are already present in the well, a special diverter is installed at the end of the pipe. This device redirects the cement slurry flow in a radial direction so that no downward flow of cement slurry is formed and it is laid in a kind of "layers." There are many designs of this tool for both drill pipe and coiled tubing.

7.1.2.4 Mechanical Separators

Various mechanical separation devices (plugs, balls, etc.) are widely used to create artificial barriers between injected fluids and prevent their mixing during injection.

7.1.3 Slurry Design

In general, the process of designing a cement slurry composition for a cement plug installation practically does not differ from the procedures and recommendations, which are followed by specialists when designing compositions for primary cementing operations. However, there are some characteristic features and differences in the choice of formulation, which should be mentioned. For example, in spite of identical requirements to the thickening time (i.e. it should be sufficient to perform the operation), the rate of cement hardening should be much higher. Serious attention must also be paid to such parameters as the segregation of the cement mortar and water separation. A high density of the composition used is also preferable. The exception is when the cement bridge is installed in the interval of weakly cemented rocks and high density of mortar may lead to hydraulic fracturing of the formation. Cement slurry shrinkage values should also be minimal.

If the purpose of plug cementing is well abandonment, cement stone should have low permeability, preferably less than 0.001 mD. If installing a cement plug in an open wellbore, cement slurry fluid loss values must not exceed allowable standards.

When installing a cement plug for directional drilling, the strength of the cement slurry must be greater than the strength of the surrounding formation. It is not uncommon for these values to exceed the possible strength of the cement stone and operators add reinforcing elements to the cement slurry.

Often the cement system used must be resistant to subsequent possible stimulation operations. For example, be highly resistant to acids commonly used for this purpose. Latex or epoxy-based cement slurries are the most commonly used solutions in this case.

An equally important factor is the bonding strength between the cement stone and the casing because it is what provides the necessary durability of the cement plug and allows it to withstand the subsequent high mechanical stresses encountered during production operations. The bonding strength of cement to steel is usually assumed to be a small fraction of the compressive strength of the cement. This value can be increased by using expanding cement systems or shrinkage-free materials (e.g. silicone cement systems).

7.1.4 Plug Cementing Evaluation

The analysis and evaluation of a successful cement plug installation are performed after the WoC (Wait on Cement) time has elapsed and the procedure is fairly straightforward:

- The cement plug installed for directional drilling is tested for its ability to withstand mechanical loads during sidetracking
- The effectiveness of a cement plug installed to prevent loss of circulation is evaluated by comparing the loss rates before and after treatment.
- The criterion for evaluating the quality of a cement plug installed for well abandonment is its permeability, which is evaluated by monitoring fluid levels in the well

If the goal of the job has not been achieved or the problem has not been corrected, the reasons for the failure should be thoroughly investigated and appropriate changes made. As noted earlier, the most common causes of failure are due to contamination of the cement slurry or insufficient volume.

7.2 Squeeze Cementing

As mentioned earlier, squeeze cementing has many applications.

- Repairing primary cementing defects
- Water shut-off
- Reducing the gas and water cut of the produced fluid
- Fixing casing leaks caused by corrosion or casing fracture
- Isolation of depleted or unproductive reservoir intervals

- Changing the injectivity profile of the well
- Isolation of lost circulation zones
- Isolation of behind-the-casing flows.

Squeeze cementing operations implies pumping of cement slurry under pressure much higher than hydrostatic pressure since cement slurry is a suspension, poorly filtered through a porous medium. As a result, solid particles accumulate at the cement slurry/porous media interface, forming a filtration crust, while filtrate penetrates further into the formation. The filter cake increases in size and forms small protrusions in the wellbore as it is injected under pressure (Figure 7.4).

The formed filtration crust is conventionally considered incompressible. To determine the time required to create a filtration crust of a given height h_{fc} at a constant differential pressure Δp, usually applies Darcy's law, assuming that the differential pressure is constant over the entire area of the filtration crust. The equation for predicting the thickness of cement slurry filtration crust (h_{fc}), adapted to the parameters of the experiment for measuring fluid loss according to API/ISO standards, is as follows:

$$h_{fc} = \omega \times \frac{V_{API}}{A_{API}} \times \sqrt{\frac{\Delta P}{\Delta P_{API}}} \times \sqrt{\frac{t}{t_{API}}} \qquad (7.1)$$

$$\omega = \frac{f_{sV}}{1 - f_{sV} - \varphi} \qquad (7.2)$$

$$f_{sV} = \frac{V_{solid}}{V_{slurry}} \qquad (7.3)$$

where ω – deposition factor, f_{sV} – solid volume fraction, $A_{API} = 3.5\,\text{in.}^2$ (filtration mesh area), ΔP – Filtration pressure, $\Delta P_{API} = 1000\,\text{psi}$ (differential pressure at which the fluid loss of cement slurry is determined), $t_{API} = 30\,\text{min}$ (the duration of the fluid loss test), t – the time required for the formation of a filter cake, V_{API} – filtrate volume, V_{solid} – volume of solids, V_{slurry} – volume of slurry, and φ – porosity

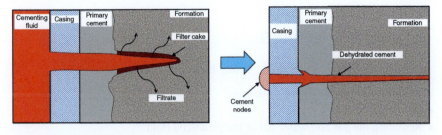

Figure 7.4 Schematic representation of the formation of a filtration crust and cement nodes in the wellbore.

of the filter cake. By converting Formula (7.1), we can calculate the time required for the formation of the filtration crust:

$$t \approx \frac{6.1 \times 10^7}{\Delta P} \left(\frac{h_{fc}}{\omega V_{API}} \right)^2 \tag{7.4}$$

Using Formula (7.4), calculate the time it takes for a 2-in. (5.1 cm) thick filter cake to form a slurry with an API/ISO 80 ml/30 min fluid loss at 500 psi (3.5 MPa) pressure drop:

$$t \approx \frac{6.1 * 10^7}{500} \left(\frac{5}{1.3 \times 80} \right) \approx 180 \text{ min}$$

It is important to note that these formulas provide only a rough understanding of the thickness of the filtration crust and the time of its formation. Care should be taken to extrapolate these results to actual downhole conditions.

Many experts accept a 2-in. (5.1 cm) thickness as acceptable for most squeeze cementing operations and stable filter cakes. Figure 7.5 is a schematic representation of the resulting wellbore protrusions as a function of cement fluid loss.

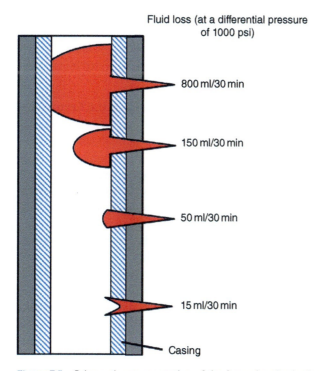

Figure 7.5 Schematic representation of the formed nodes in the wellbore depending on the fluid loss of the cement slurry.

The generally accepted rules are: the use of low fluid loss rates for operations in highly permeable formations and the use of cement solutions with high fluid loss rates in low-permeable formations. In essence, both recommendations are aimed at reducing the time of filtration crust formation.

7.2.1 Squeeze Cementing Technologies

Squeeze cementing technologies are classified according to three main criteria:

- According to the squeezing pressure:
  - squeezing pressure is lower than the formation hydraulic fracture pressure
  - squeezing pressure is higher than the formation hydraulic fracture pressure
- By the way of injecting the cement slurry:
  - continuous injection of cement slurry
  - discontinuous injection of cement slurry
- How the operation is carried out:
  - without a cementing packer
  - with a cementing packer

7.2.1.1 Classification of Squeeze Cementing Technologies According to Squeezing Pressure

The squeezing pressure is the pressure at which the cementing job is carried out. According to this criterion, squeeze cementing operations are divided into the following two groups:

- **Operations performed at a squeeze pressure lower than the formation's hydraulic fracture pressure.** One of the most critical factors for a successful cementing technique is the absence of drilling cuttings or other contaminating material in the zones of injection of the cement slurry (perforation holes, fractures, etc.). In case the operation is carried out on a producing well, such complications, as a rule, do not arise, but in the wells recently commissioned, it is often required to carry out operations on cleaning of cement slurry injection zones. Most specialists consider this technology as the most effective and preferable.
- **Operations performed at injection pressure higher than formation hydraulic fracture pressure.** Despite the undesirability of exceeding the pressure of formation hydraulic fracturing, unfortunately, in some cases, it is practically unavoidable. Examples are the following most common scenarios:
  - Channels or fractures in the annular space have no direct communication with the perforation holes.
  - Small size of cracks and channels in annular space, freely permeable for gas, but practically not permeable for cement slurry.
  - The presence of material clogging the fractures or channels, which cannot be removed by traditional hole-cleaning methods.

In all these scenarios, the only solution is to create pressure in excess of reservoir pressure, thereby inducing hydraulic fracturing. As a result, high-permeability channels and fractures are created that allow pumping of cement slurry to the desired zone, and in the case of clogging material, squeezing it further into the formation, thereby freeing up space for the cement slurry. It should be noted that the size and direction of cracks in such operations cannot be controlled, which is one of the most serious drawbacks of this technology.

7.2.1.2 Classification of Squeeze Cementing Technologies Depending on the Method of Injection of Cement Slurry

There are two types of squeeze cementing technologies according to the method of injection of cement slurry:

- **Continuous injection of cement slurry.** In this method, the cement slurry is pumped continuously until a certain pressure is reached, which may be higher or lower than fracturing pressure (Figure 7.6). The operation is considered successful and finished if after pumping the calculated volume of cementing solution, the pressure indicators remain unchanged for several minutes. Otherwise, an additional amount of cement slurry is pumped until stable pressure values are achieved. As a rule, the volume of injected cement slurry in this method is rather large.

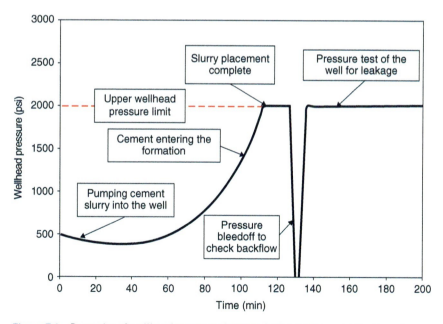

Figure 7.6 Dynamics of wellhead pressure changes during continuous injection of cementing slurry.

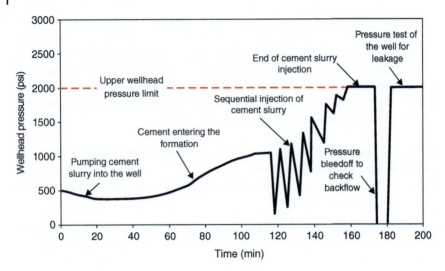

Figure 7.7 Wellhead pressure changes during sequential injection of cementing slurry.

- **Discontinuous injection of cement slurry.** Technical capabilities of equipment used for injection of cementing solution do not allow to decrease injection rate to values comparable with the rate of filtration of cement slurry into formation, which eventually negates all attempts to keep differential pressure constant, especially at injection pressure exceeding hydraulic fracturing pressure. The solution to the problem is the discontinuous injection of cement slurry, which consists in injecting small volumes of cement slurry, separated by waiting intervals of 10–20 minutes. When the first packs of cement are being pumped, as the filtration crust is not yet formed, the filtrate leaks out quickly enough. At the end of 10–20 minutes' interval, the pressure in the well significantly decreases in comparison to the time when the pumping was stopped. However, as the filtration crust is formed, the leakage rate decreases, and this difference also begins to decrease (Figure 7.7). This method requires small volumes of cementing fluid. The wait interval value depends on the permeability of the target formation. As permeability values increase, the wait interval also increases.

7.2.1.3 Classification of Squeeze Cementing Technologies According to the Method of Operation

There are two types of squeeze cementing technologies based on the way the operation is performed:

- **Without using a cementing packer.** This method does not require any additional equipment and is used in situations when the well construction allows

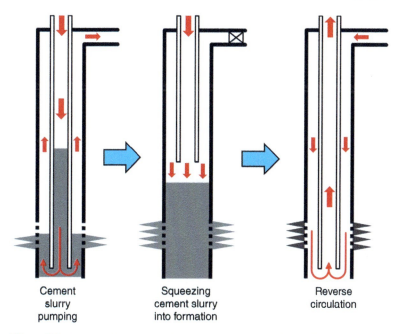

Figure 7.8 Schematic representation of squeeze cementing without a packer.

withstanding the squeeze pressure (Figure 7.8). The exception is when it is necessary to isolate downhole intervals, which are usually plugged with a plug bridge. This method is technically simple and therefore is quite widespread.
- **With the use of a cementing packer.** This method is subdivided, depending on the type of packer used, into the drillable packer method and the retrievable packer method.

7.2.2 Slurry Design

The composition of cement slurry during squeeze cementing operations should be selected in such a way that the cement slurry should perform the following functions:

- Possibility of pumping a cement slurry into bottomhole zone.
- Possibility to inject cement slurry into annular space or perforation holes.
- Provision of required physical and mechanical properties of hardened cement stone.

Nowadays, quite a wide range of cementing slurries is used for operations with pressure: from compositions based on the usual Portland cement to complex cementing systems based on polymers, resins, and so on. Nevertheless, in spite of the variety of cementing systems used, specialists single out a number of general recommendations, which they have to comply with:

- low viscosity values, allowing the solution to penetrate into small fractures,
- low values of gel strength – as the transition into the gel state significantly increases hydraulic resistance of the cement flow, which, in turn, leads to higher wellhead pressure, which is difficult to interpret correctly,
- cement particles must have an appropriate particle size distribution,
- free water must be kept to a minimum and preferably none at all,
- the slurry must have the appropriate fluid loss values,
- the thickening time of the cement slurry should be long enough to ensure safe performance of the job.

Let us take a closer look at a number of critical physical and chemical properties and parameters when developing a cement slurry formulation.

7.2.2.1 Fluid Loss

The requirements for cement slurry fluid loss values are directly related to formation properties and the type of voids or fractures to be filled. Typically, engineers are guided by the results of an API/ISO slurry water release test when selecting this parameter.

One approach is to select cement slurry fluid loss values depending on the permeability of the formation. Thus, for formations with permeability below 100 mD, it is preferable to use compositions with low fluid loss values – less than 100 ml/30 min. In case the purpose of the operation is to fill in shallow annular channels or microcracks, it is not uncommon to use solutions with fluid loss values less than 50 ml/30 min.

In highly permeable formations, as a rule, the compositions with high fluid loss values of 300 to 500 ml/30 min are used, in order to provide quick formation of filtration crust. It is also possible to use two cement slurries in series, when initially a certain volume of cement slurry with high fluid loss values is pumped, which allows to quickly create filtration crust, followed by slurry with lower values of this parameter.

It should be noted that, despite the fact that the formation of filtration crust is one of the key factors of successful squeeze cementing operation, it is important to control this parameter very carefully. If the filtration crust generation rate is too low for a given formation, the chance of successful operation will significantly decrease, and if the filtration crust generation rate is high, cement protrusions inside the casing may significantly reduce casing inner diameter, thus complicating well operation.

7.2.2.2 Rheology

The regulation of rheological properties of cement slurries used in squeeze cementing is probably one of the most difficult tasks for an engineer because the requirements for these parameters often contradict each other. For example, filtration properties of slurries directly depend on viscosity values. High-viscosity cement slurries are very effective in filling large voids and cracks, but are practically useless in the case of cracks and voids of small size, as they will not penetrate them. Of course, this complication can be solved by increasing the injection pressure to higher than the fracturing pressure, but this is not always an acceptable way out of the situation. Dispersants are often a much more suitable solution. However, too low values of cement slurry viscosity can also have negative consequences in the form of the formation of free water, sludge formation, or segregation.

As mentioned earlier, high gel strength values are also a negative factor since piezoconductivity and, consequently, the formation of a filter cake are impaired. But again, having high gel strengths is a good way to prevent the cement slurry from sinking in the annulus if there is a lighter fluid or no mechanical barrier underneath.

7.2.2.3 Thickening Time

As with primary cementing, temperature and pressure are important factors affecting slurry thickening time. The duration of the cement slurry thickening time must be long enough to support cement injection, squeezing, and hole-cleaning operations. Temperatures encountered during cement slurry squeezing may be somewhat higher than during primary cementing operations because the circulation process takes much less time and, consequently, the well is less cooled. The API standards separately prescribe testing of cement slurries used in squeeze cementing.

7.2.3 Design and Execution of Squeeze Cementing Operations

7.2.3.1 Determination of the Cement Slurry Volume

The optimal volume of cement slurry as a rule is determined on the basis of previous experience of works and general empirical recommendations. There is no strict methodology for solving this issue. Let us take a closer look at some generally accepted recommendations for selecting the volume of cement for squeeze cementing operations.

A good practice for the selection of the necessary volume of cementing slurry is the initial determination of the volume of voids and fractures by carrying out injectivity tests of the well. When performing an operation below fracture pressure, a significantly smaller volume of cementing fluid is required, in many cases, less than one barrel. However, a batch of 5–15 barrels is often prepared for ease of operation.

High-pressure squeezing, where formation fracturing occurs, requires a larger volume of slurry, depending on the geometric dimensions of the fractures being created. Mud consumption can be minimized by operating at low pump speeds comparable to fracture generation rates. Otherwise, the volume of cement slurry entering the formation will not be enough to fill the formed fractures and the formation of a filtering cake, which will lead to increased consumption of cement slurry.

It is worth noting, of course, a number of limitations when selecting volumes of cement slurry. So, for example, it is necessary to strictly control the hydrostatic pressure indicators in case of using large volumes of cement slurry; otherwise, it is quite realistic not to plan the hydrofracturing of poorly cemented formations. The volume of cement slurry also should not exceed the capacity of casing, and in case of application of coiled tubing, the weight of the cement-filled tube should not exceed the limit of tensile strength of the tube.

7.2.3.2 Spacer, Washer, and Displacing Fluids

Spacer and washer fluids have two main functions in squeeze cementing:

- Prevent contamination of the cement slurry.
- Clean the voids and perforations that are to be filled with cement.

As mentioned earlier, cement slurry contamination is one of the most serious risk factors in squeeze cementing operations. The volumes of cement slurry are often not large and even minor contamination of cement slurry by primary cementing standards can cause a number of serious complications associated with changes in slurry thickening time, fluid loss, and viscosity. Spacer and washer fluids are used to prevent such a negative scenario. It is also possible to use special devices for flushing perforation holes, which are usually used together with the pumping of various acids. By the way, there is an additional positive effect from the use of acids: they weaken the structure of the formation and thus reduce the values of hydraulic fracturing pressure. Mechanical separators are the most effective tools to prevent the mixing of cement with other fluids in the production casing, but in the annulus, this role can only be performed by spacer fluids.

As a rule, the volume of displacement fluid corresponds to the inner volume of the production casing from the wellhead to the top perforations to be cemented. In some cases, a few more barrels are added to this volume for safety reasons. However, the maximum volume of the displacement fluid may be uncertain due to factors such as pump efficiency, variability in internal casing volume, and compressibility of the fluids. The latter parameter is most important because the pressures during a squeezing operation are much higher than during primary cementing operations. When performing the operation in an open-hole section, there may be some compression of the formation, which will increase the volume of the well. In the case of a nonpacker operation, it is also possible to expand the casing itself, which will also increase the required displacement fluid volume.

7.2.3.3 Determination of Well Injectivity

Before injecting the cement slurry, the well is tested for injectivity. The results of this test allow to make sure that the perforations are not plugged, to predict such parameters as volume, pressure, and rate of injection of the cement slurry. For injectivity tests, water is usually used as the working fluid. If water cannot be pumped, acid treatment with hydrochloric or hydrofluoric acid is used. As mentioned before, one of the goals of injectivity tests is to collect data for predicting pressures during pressurized cementing. For this purpose, water is injected into perforation holes for several minutes at a flow rate close to that predicted for the job until a pseudo-steady-state condition is reached. At the same time, the injection pressure should be maintained below the fracturing pressure. In general, as the injectivity of the well decreases, the importance of the properly selected composition of the cementing solution increases.

The volume of voids in the annulus is determined somewhat differently. For this purpose, a small volume of high-viscosity fluid is injected. The viscosity of the injected fluid is selected so as to create a measurable increase in pressure at the wellhead. The volume of voids is assumed to be approximately equal to the volume of high-viscosity fluid injected during the pressure increase. After the test is completed, the high-viscosity fluid must be removed to avoid problems caused by incompatibility with the cement slurry.

7.2.3.4 Main Procedures for Squeeze Cementing Operations

The following describes the general sequence of operations during squeeze cementing operations:

1) Isolate areas below the interval of operations, by means of a retrievable or drillable bridge plug.
2) Flush the perforations with an appropriate tool or reopen them.
3) Conduct a borehole injectivity test.
4) Circulate the fluid used in the injectivity test and then the cement slurry in the well, with the packer bypass open, to prevent the fluids in the well or annulus from being forced into the formation before the cement slurry. A small back pressure is created in the annulus to prevent the free fall of the mud as a result of the U-tube effect.

5) Increase pressure on operating values
6) Pump in cement slurry until stable pressure values are reached. Completion is indicated by an increase in pressure values of 500 psi (3.5 MPa) greater than the final injection pressure.
7) Pressure is released. The packer bypass is opened and excess cement is evacuated in the opposite direction.
8) The operating casing is retrieved and the WoC (Wait on Cement) time is maintained.

7.2.4 Analysis and Evaluation of the Squeeze Cementing Job

When evaluating the results of squeeze cementing operations, it is necessary, first of all, to take into account the requirements for subsequent planned operations in the well. As the range of such operations is quite wide, the evaluation of the success of operations is carried out individually for each individual case. However, regardless of the goals and objectives to be achieved, in almost all cases, the borehole condition is evaluated for excess cement in the casing and intra-casing cement protrusions that reduce the inner diameter of the casing. One of the most widespread methods of works assessment is a temperature log, which allows determining the location of cement slurry in the annulus.

Another widespread method is the pressure test, which is based on creating positive and negative pressure in the borehole. The positive pressure test increases the pressure in the borehole to the design values and determines the injectivity of the well.

The negative pressure test (inflow test) consists of reducing downhole pressure and is used to determine how effectively plugged perforations prevent fluid from entering the well. The pressure in this test usually corresponds to the pressure the well is expected to experience during operation and is determined during the design phase of the work.

The process consists of placing low-density brine against the perforations, followed by swabbing the well and flow testing the perforations (Figure 7.9).

If the perforations are completely tight, the test pressure graph (Figure 7.10) should not show fluid flowing from the well. However, the test pressure must be strictly controlled, otherwise, if the pressure is reduced excessively, the cement stone and, as a consequence, the perforations may collapse.

In case of squeeze cementing to correct primary cementing defects, acoustic logging data performed before and after completion of the work may be a good tool for evaluating the quality of the work performed. Also, a comparison of well flow rates before and after the work can be used for this purpose. To assess the placement of cement in the annulus, a possible solution is also the use of radioactive tracers.

Figure 7.9 Schematic representation of the negative pressure test (inflow test).

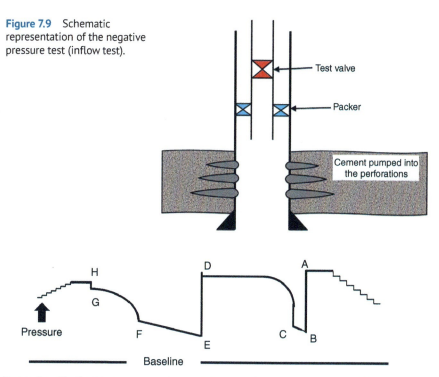

Dynamics of hydrostatic pressure change before the start of squeeze cementing operation
- A   Initial hydrostatic pressure with packer installed
- B   Initial hydrostatic pressure at the opening of the test valve
- B-C Pumping of cement slurry
- C   Test valve closed
- C-D Recovery of hydrostatic pressure
- D   Initial hydrostatic pressure
- E   Cement injection pressure after second opening of the test valve
- E-F Pumping of cement slurry
- F   Test valve closed
- G   Wellhead pressure after reclosing the test valve
- H   Final hydrostatic pressure after removal of the packer

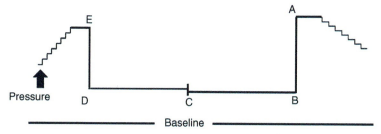

Dynamics of hydrostatic pressure change after successful squeeze cementing operation
- A   Initial hydrostatic pressure with packer installed
- B   Hydrostatic pressure when the test valve is opened
- C   Hydrostatic pressure when the test valve is closed
- D   Hydrostatic pressure at the end of the shut-in period
- E   Final hydrostatic pressure after removal of the packer

Figure 7.10 Dynamics of hydrostatic pressure change during a negative pressure test.

8

Cement Job Evaluation

The task of casing cementing assessment is to evaluate the success of cementing operations in terms of goals and tasks, depending on the type and purpose of cementing operations. Though the main task of all cementing jobs is to ensure proper casing isolation, depending on the type and purpose of casing, cementing jobs have different specific tasks, and a precise understanding of these tasks is very important, otherwise the process of job evaluation and assessment would be useless. For example, when cementing surface casing, the main objective is to prevent erosion processes in annular space, if we are talking about cementing the first intermediate casing, the purpose of cementing operations is to isolate water-saturated formations and provide a solid foundation for the second intermediate casing. The second intermediate casing is cemented to isolate intervals with abnormal reservoir pressure and to close zones of lost circulation, while the production casing is cemented to prevent fluid migration in the annulus and to provide zonal isolation. Regardless of the type of casing to be cemented, the cement also provides some corrosion protection for the casing itself.

In remedial cementing operations, the main goals and objectives of the operations performed are to correct primary cementing defects, repair casing leaks, and isolate productive formations.

Before the advent of modern logging technologies, quality assessment of cementing operations consisted of determining the height of cement uplift in the annulus and evaluating the quality of hydraulic isolation. However, evaluating the quality of hydraulic isolation often required casing perforation or additional drilling operations. Certainly, in situations where the purpose of cementing operations was only to create a cement sheath around the casing, cement lift height data are sufficient to assess the success of the operation, but more sophisticated evaluation methods and tools must be used to verify the quality of the interzone isolation.

This chapter presents the main methods and tools for evaluating the success of a cementing operation to date.

Oil and Gas Well Cementing for Engineers, First Edition. Baghir A. Suleimanov, Elchin F. Veliyev, and Azizagha A. Aliyev.
© 2023 John Wiley & Sons Ltd. Published 2023 by John Wiley & Sons Ltd.

8.1 Hydraulic Testing

Hydraulic tests are methods of assessing the isolation provided by the cement, both after primary cement works and after remedial cementing. The most common methods are hydraulic pressure test (pressure test) and inflow test.

8.1.1 Pressure Test

Pressure testing is the most common method of hydraulic testing. It is usually carried out after each cementing of the intermediate casing. Initially, before the cement sets, the casing is pressurized to check its mechanical integrity. To avoid damage to the cement that has been set or to the cement bond with the casing, this stage of testing should be carried out immediately after the top cement plug is seated.

In the second stage, after the cement has set, the casing shoe is drilled and the internal pressure in the casing is increased to values higher than those of the subsequent drilling phases. The objective of this stage of the test is to assess the tightness of the lowest cementing interval and, if the pressure in this interval decreases, i.e. if there is a pressure leak, remedial cementing is performed. In case the pressure can be raised to the values of hydraulic fracture, the test is called pressure-integrity test (PIT) or leak-off test. The purpose of this test is to determine the maximum possible density of the drilling fluid that can be used for subsequent intervals. The leak-off test is most commonly performed on exploration wells. The general procedure for the test is as follows:

1) Preparation for the test. This includes checking the equipment being used for leaks, calibrating gauges, and preparing a test fluid (e.g. drilling fluid).
2) Pressurizing the casing.
3) Preparing test schedule.
4) Pumping in drilling mud at a constant rate, 0.25–1 barrel per minute and plotting the pressure readings on the test graph.
5) Determine and correct the pump flow rate values at the maximum volume line.
6) If the curve deviates from a linear trend, a small additional volume of drilling fluid is injected and pumping stopped.
7) The pressure drop is monitored for 10–15 minutes.

This test plot reveals various patterns, in particular the presence of a channel in the cement may be indicated by any of the following (Figure 8.1):

- Equivalent mud weight is below the predicted value.
- The shut-off pressure is not equalized.

8.1 Hydraulic Testing

(a) Dynamics of pressure changes depending on the volume of injected working agent

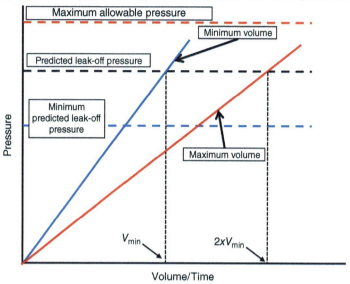

(b) Dynamics of pressure changes in the process of preparation and performance of the leak test

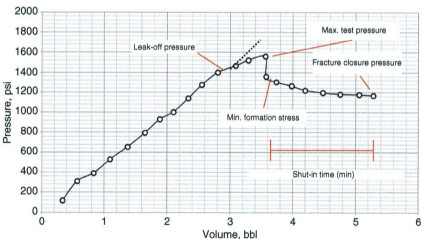

Figure 8.1 Leak-off test.

(c) Example interpretation of the pressure dynamics in the presence of leakage behind the casing

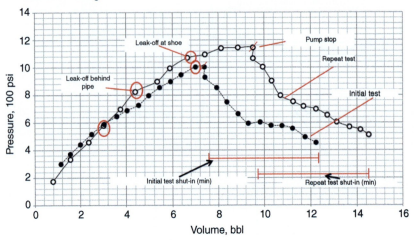

(d) Pressure changes after the leak-off test

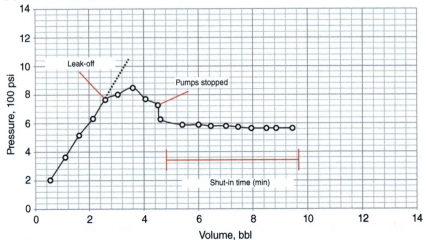

Figure 8.1 (Continued)

If these indicators are unchanged and the tests are repeated, the presence of a channel in the cement stone can be considered confirmed. In the case of deep well tests, it is not uncommon to make certain corrections to the results. Such corrections usually include compressibility of the fluid, thermal expansion, friction losses, etc. For shallow wells, no such correlation of the results is usually carried out.

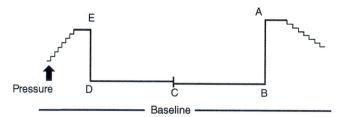

Dynamics of hydrostatic pressure change after successful squeeze cementing operation

A   Initial hydrostatic pressure with packer installed
B   Hydrostatic pressure when the test valve is opened
C   Hydrostatic pressure when the test valve is closed
D   Hydrostatic pressure at the end of the shut-in period
E   Final hydrostatic pressure after removal of the packer

Figure 8.2 Dynamics of hydrostatic pressure change during negative pressure test (inflow test).

8.1.2 Inflow Test

The inflow test (dry test) is a drill stem tester (DST) run to assess the isolation provided by the cement. In essence, this test is the exact opposite of a hydraulic pressure test. The pressure inside the casing is lowered and the well is monitored for the flow of formation fluids. A successful test is one in which there is no change in bottomhole pressure while the bottomhole valve is open or during a production shut-in (Figure 8.2). The flow test is particularly useful to verify the effectiveness of pressure cementing or to evaluate the quality of isolation at the top of the casing.

8.2 Temperature Log

Temperature logging is often used to assess primary cement work, mainly for cement uplift in the annulus, detection of leakage in the cement sheath, and channel creation. The measurements are carried out using a fiber optic cable inside the annulus (in the control line) or inside the casing. The temperature change over time and depth is recorded, allowing the presence of cement in the annulus to be accurately determined. Temperature logging is typically carried out a few hours after cementing operations have been completed. As was described earlier, chemical reactions during the hydration of cement are exothermic, i.e. accompanied by heat release. As a result, wellbore temperature increases and a deviation from the normal temperature gradient is observed. Figure 8.3 shows the results of the

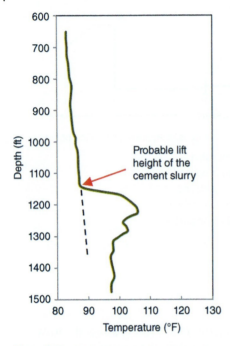

Figure 8.3 Typical temperature log curve.

temperature log. The maximum temperature anomalies occurring during cement hydration range from 18 °C to 70 °C and are highly dependent on the geometry of the borehole space and the thermal conductivity of the formation. The amount of heat released during cement hydration is also closely related to the amount of cement and additives present in the cement slurry. Compared to standard cements, low-density cements containing extenders generate less heat per unit volume and the detection of temperature anomalies can be difficult.

Well cooling due to the circulation of working fluids before and during cementing also affects cement hydration kinetics – the longer the circulation, the lower the temperature. The cooling of the well also prolongs cement slurry thickening time, so that hydration reactions occur later, which must be taken into account for temperature logging. This type of measurement may not be suitable for evaluating very long cementing intervals, as the temperature difference between the upper and lower intervals may be significant, and the cement at the top of the string may take longer to set.

The borehole temperature also plays an important role, as the temperature anomaly readings for the same mortar will be significantly higher at higher curing temperatures (i.e. deeper or high temperature wells) (Figure 8.4). In most cases, the temperature deviation peaks after 4–12 hours and persists for a further 24 hours. A general recommendation is to conduct a temperature log during the first 12–24 hours after cementing operations are completed. Naturally, the amount of heat generated is directly related to the volume of cement and, as a consequence, will be greater in large annular spaces. In narrow annular spaces, the amount of heat released during cement hydration may not be sufficient to significantly change the temperature profile in the wellbore and provide a good temperature log.

Temperature-logging data can be an effective tool for assessing the presence of channels in the cement rock. To do so, a defined volume of working fluid is

Figure 8.4 Influence of temperature on the hydration kinetics of G-class cement.

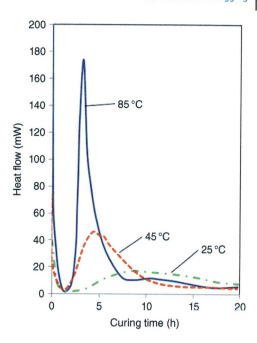

injected into the annulus and the results of the temperature log before and after the operation are compared. The presence of a significant temperature drop above the perforation area and temperature fluctuations up to the oil-water contact can be good indicators of the presence of overflows in the annulus.

8.3 Radioactive Logging

Radioactive materials are often used in the oil industry as tracers. This method is widely used to determine the location of drilling fluids, the timing of their circulation, and determine penetration in various stimulation operations. In evaluating the success of cementing operations, this method is used to determine the height of cement rise in the annulus (Figure 8.5). It is important to emphasize that special health and safety measures must be taken when using radioactive materials, especially those with a long half-life. The radioactive logging method consists of comparing the spectral gamma ray (GR) log before and after cementing. A calculated concentration of radioactive material is added to the cement slurry beforehand.

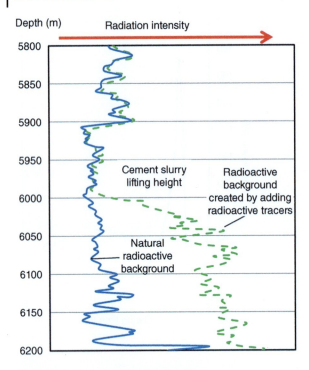

Figure 8.5 Typical radioactive log diagram.

Several types of radioactive tracers have been used to date. Soluble tracers (e.g. I131) are added directly to the mixing water. Bulk materials coated with radioactive material (e.g. Ir192) are usually added to dry cement or cement mix.

The main criterion for the selection of a radioactive tracer is its half-life. With a long half-life, changes in the initial GR signature in the formation will be constant over an extended period of time and this must be taken into account in the case of subsequent radioactive logging. With a short half-life, the tracer will cease to be radioactive within a few weeks or months, and the GR signature will return to its original state. A second selection criterion is the energy of the dominant GRs emitted by the tracer. When spectral GR logging is used, the radioactivity of the tracer can be selectively measured and the amount of radioactive material required for logging can be significantly reduced. Table 8.1 lists the most common radioactive tracers with information on their half-lives and GR energy.

8.3.1 Pulsed Neutron Logging

The pulsed neutron logging procedure is similar to the temperature logging procedure with working fluid injected into the annulus. Boron is often used as a radioactive tracer. A boron solution is injected through a perforation and pulsed

Table 8.1 Radioactive tracers.

Name	Half-life (day)	Gamma radiation energy (MeV)
Cr^{51}	27.7	0.32
Fe^{59}	45	1.10
Br^{82}	1.5	0.77
I^{131}	8	0.36
Ir^{192}	74	0.32
Au^{196}	2.7	0.41

neutron logging is carried out. The boron absorbs neutrons very effectively so that channels in the cement sheath can be easily detected even through two casing strings. This method is more accurate than temperature logging.

8.3.1.1 Oxygen-Activated Neutron Gamma Method

This measurement method allows a quantitative assessment of the presence of water-saturated channels in the annulus. High-energy neutrons interact with oxygen nuclei in the water, resulting in the emission of GRs. Before the borehole is charged with neutrons, an initial spectral GR log is constructed. The neutron tool is then activated for a very short period (1–15 seconds), followed by a longer period of GR registration (20–60 seconds). The measurement data allows the water flow rate in the borehole, its location, and volume to be determined.

8.4 Acoustic Logging

Acoustic logging is the most widely used and effective method for evaluating cementing jobs. This measurement method is based on the acoustic properties of the environment (casing, cement, and formation) and the quality of the acoustic coupling between the casing, cement, and formation. The presence of cement is only one of many parameters that can affect the acoustic properties of the media. So, it is important to interpret the results carefully. In most cases, detailed information about well geometry, formation characteristics, and cementing operations is required. Proper analysis of measurement results is only possible by comparing expected results with actual data. Today, it is possible to classify and possibly quantify cementing results mainly in terms of cementing stone quality and the height of its rise in the annulus.

In all well logs, it is necessary to have so-called "re-recorded sections." Re-recorded sections of well logs are diagrams of small intervals of the well that are part of the whole measurement interval, usually not exceeding 60 m, recorded before the start of the main measurements. Both measurements must be taken under the same conditions, including tool settings and borehole conditions. The purpose of this approach is to ensure that the logging tool repeatedly gives the same readings under the same conditions. However, it is worth noting that even when results are repeated, there is no 100% guarantee that the results are accurate.

The standard of "good repeatability" depends on the type of tool, how it works, and its design. Unlike the tools used in radioactive logging, the tools used in acoustic logging do not experience statistical fluctuations, and the repeatability of the results obtained depends primarily on the quality of the tool design.

To maximize acoustic coupling between the casing and cement, some operators increase the pressure in the casing during logging. The sensitivity of acoustic measurements to pressure changes requires building and maintaining the same pressure throughout repeat and main measurements. Depending on the equipment used to build pressure and maintain casing integrity, this scenario is not always possible. In such a case, the only way to ensure repeatability of the results obtained would be to conduct logging without additional internal pressure in the casing.

It should also be noted that there is currently no industry standard for calibrating acoustic logging tools, and it is more a test of tool performance than a calibration.

Since acoustic logging is based on the propagation of sound waves, let us briefly consider some important parameters for understanding the process.

The study of sound is known as acoustics. It refers to the propagation of sound waves in the context of acoustic logging. The periodic rarefaction and compression of molecules (in the case of a gas or liquid) or the squeezing and stretching of particles (in the case of a solid) is what causes sound to travel (Figure 8.6). When these

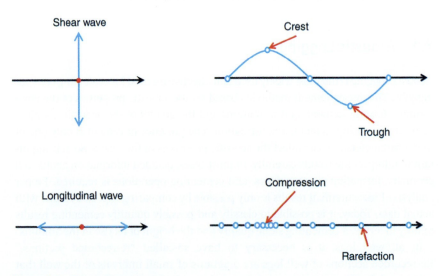

Figure 8.6 Types of sound waves.

vibrations occur in the same direction as the wave propagates, they are called compression waves. In a solid, a second type of wave is called a shear wave. In shear waves, the oscillations are perpendicular to the direction of wave propagation. In liquids, no such wave is formed and it is only inherent to solids. The speed of shear waves is always less than the speed of compression waves. The velocities of both types of waves depend on the elastic properties (i.e. Young's modulus, shear modulus, and Poisson's ratio) of the materials through which they pass and are almost independent of the wave frequency.

Another type of wave important for acoustic logging is the Lamb wave – a complex shear wave propagating in an elastic medium (Figure 8.7).

This type of wave propagates along the steel casing at a slightly slower rate than shear waves and is one of the main parameters measured in acoustic logging.

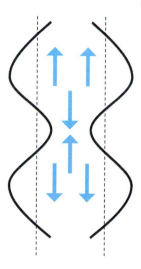

Figure 8.7 Lamb wave.

The acoustic properties of the formation also influence acoustic logging results. Depending on the sound velocity, there are two types of formations: high-velocity and low-velocity. In acoustic logging after cementing, a formation is traditionally called a high-speed formation if the sound travels through it faster than along the casing (i.e. 17 000 ft/s [5334 m/s]). An equally important parameter is sound attenuation, that is, the decrease in the intensity of the sound wave as it propagates; it is quite large in unconsolidated formations and negligible in strong, consolidated rocks.

The acoustic properties of rocks are well known and studied, and they remain virtually unchanged throughout the operating life of a well. In contrast, the acoustic properties of cement change over this period of time. This fact should be taken into account when conducting acoustic logging because changes in the physical properties of cement over time will lead to different acoustic logging results and may lead to misinterpretation of the results obtained.

The acoustic properties of different cement compositions are presented in Table 8.2. Thus, low-density cement compositions have a low acoustic impedance, which can change significantly after a few days. The acoustic impedance of higher-density cement compositions changes by less than 20% in a period of one to seven days. Acoustic impedance is a complex acoustic resistance of the medium, which is the ratio of sound pressure to effective particle velocity.

Light cement solutions with additives of microspheres or nitrogen have a very low acoustic impedance, which significantly complicates the interpretation of the logging results.

The exact knowledge of the acoustic properties of fluids both in the casing and in the annulus is also critical for acoustic logging. Table 8.3 lists the

Table 8.2 Acoustic properties of different cementing formulations.

Composition of cement slurry	Density (ppg (kg/m³))	Time (day)	Speed of sound propagation in cement (m/s)	Acoustic impedance (MRayl)	Change in acoustic impedance after 24 h (%)
Class G neat cement	15.8 (1.89)	1	3000	5.68	0
		2	3250	6.16	8
		7	3400	6.44	13
Class G cement + latex + hollow silica gel microspheres	11.2 (1.34)	1	1650	2.21	0
		2	2200	2.95	33
		7	2500	3.36	52
Class G cement + soluble silicate-based extender	12.0 (1.44)	1	1600	2.30	0
		2	1750	2.52	9
		7	2000	2.88	25
Class G cement + hollow silica gel microspheres + 4% $CaCl_2$ (BWOC)	12.0 (1.44)	1	2600	3.74	0
		2	2800	4.03	8
		7	3000	4.32	16
Class G cement + soluble silicate-based extender	13.3 (1.59)	1	1750	2.79	0
		2	2200	3.51	26
		7	2500	3.99	43
Class G cement + latex	15.8 (1.89)	1	2900	5.49	0
		2	3150	5.97	9
		7	3350	6.35	16
Class G cement + 18% NaCl (BWOW)	16.1 (1.93)	1	2850	5.50	0
		2	3200	6.18	12
		7	3375	6.51	18
Class G cement + hematite	19.0 (2.28)	1	3300	7.59	0
		2	3400	7.74	2
		7	3530	8.04	6
Foam cement	10.0 (1.20)	7	2300	2.76	–
Lightweight cement slurry	12.51 (1.50)	7	2000	3	–
Lightweight cement slurry with adjustable particle size	10.0 (1.20)	7	2900	3.48	–
Ultra-lightweight cement slurry with adjustable particle size	8.61 (1.03)	7	2790	2.87	–

BWOC, By Weight Of Cement; BWOW, By Weight Of Water

8.4 Acoustic Logging

Table 8.3 Acoustic properties of some homogeneous liquids.

Name	Density (ppg (kg/m^3))	Sound attenuation (µs/ft)	Sound velocity (ft/s (m/s))	Acoustic impedance (MRayl)
Water	8.33 (998)	206	4860 (1482)	1.48
Water + 10% NaCl	8.98 (1075)	193	5190 (1580)	1.70
Water + 25% NaCl	9.90 (1.186)	175	5710 (1740)	2.06
Water + 36% NaCl	11.3 (1350)	170	5870 (1790)	2.42
Water + KCl	9.18 (1100)	189	5280 (1610)	1.77
Water + 58% CaBr$_2$	15.2 (1824)	179	5580 (1700)	3.10
Seawater	8.56 (1025)	199	5020 (1531)	1.57
Kerosene	6.74 (808)	230	4340 (1324)	1.07
Diesel	7.09 (850)	221	4530 (1380)	1.17
Air at 15 psi, 32 °F (0 °C)	0.01 (1.3)	920	1090 (331)	0.0004
Air at 3000 psi, 212 °F (100 °C)	1.59 (190)	780	1280 (390)	0.1

acoustic properties of some homogeneous fluids, brines, and oils. It should also be noted that at ultrasonic frequencies, the acoustic properties of weighted drilling fluids containing weighting additives depend on inertial and viscous effects (Table 8.4).

The attenuation of ultrasound in weighted mud is sometimes a limiting factor when using acoustic logging. Attenuation increases with frequency and with increasing solid content (or density). In oil-base muds, attenuation is much faster

Table 8.4 Acoustic properties of different types of drilling mud.

Name	Density (ppg (kg/m^3))	Sound attenuation (µs/ft)	Sound velocity (ft/s (m/s))	Acoustic impedance (MRayl)	
				At low frequencies	At 0.5 MHz
Water-based drilling mud	12.6 (1510)	215	4660 (1420)	2.14	1.85
Water-based drilling mud	15.4 (1850)	216	4630 (1410)	2.60	2.14
Oil-based drilling mud	7.79 (933)	231	4330 (1320)	1.23	1.25
Oil-based drilling mud	12.6 (1510)	245	4070 (1240)	1.87	1.54

than in water-base muds. In general, attenuation is difficult to predict because it depends on mud composition, temperature, and pressure.

8.5 Types of Logging Tools

8.5.1 Cement Bond Log (CBL)

It is a basic acoustic tool that has been used in the oil and gas industry for the past 60 years. Lowered on an electrical cable, the logging tool uses a piezoelectric transducer to create a multidirectional sound wave (18–20 kHz) that travels through wellbore fluids, casing, cement, and formation. The transducers pick up the reflected waves and transmit the data to the surface.

The results are displayed as five curves, which are used to interpret cement quality and show: the GR diagram, sound transit time (TT), casing collar locator (CCL), and amplitude and variable density log (VDL).

GR log shows the natural radioactive radiation of the wellbore, distinguishing between shale and sand and limestone formations. TT is the time elapsed from the time the sonic wave is launched to the first hit in the 3-foot receiver, measured in microseconds. Casing collar locator (CCL) is a magnet that detects large volumes of casing collar metal. Amplitude is the height or strength of the first peak, measured in millivolts, also recorded on the 3-foot receiver. VDL is a time-depth curve that represents the peaks and lows of complex sound waves recorded on a 5-foot receiver (Figure 8.8).

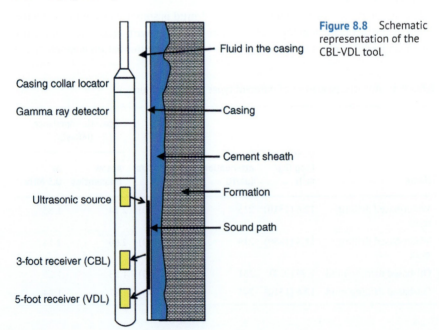

Figure 8.8 Schematic representation of the CBL-VDL tool.

8.5.2 Radial Acoustic Cement Meter

This tool is very similar to a traditional CBL tool, but in contrast, it contains an additional set of receivers that are arranged in a circle around the tool and receive signals directed to the appropriate sides of the wellbore.

The measurement results are also displayed as curves but additionally contain a so-called amplitude map (cement map) that averages the response over each measurement sector.

8.5.3 Multiple Pad Sonic Tool

This is an instrument with a scanning measurement mode, allowing measurements to be made along the perimeter of the well at (45–60)° in the radial direction. This type of tool operates at frequencies up to 100 kHz and is equipped with multiple transmitters and receivers. In comparison with CBL tools of traditional design, they allow not only detecting cementing defects but also revealing their size and spatial orientation relative to the apsidal plane of the well. Apsidal plane is a vertical plane passing through the tangent to the axis of the well at a given point.

8.5.4 Ultrasonic Tool

This instrument operates at a frequency of 200–700 kHz, and includes a rotating head that sends and receives a signal from the same transducer (Figure 8.9).

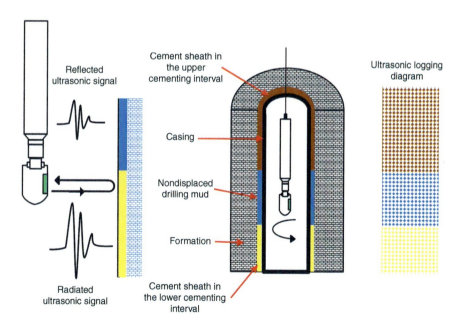

Figure 8.9 Operating principle of the acoustic logging tool.

Based on the decay rate of a part of the received signal, acoustic impedance values are recorded. Further processing of the received data helps determine casing defects and cement quality directly behind the casing. In addition to providing additional information about the casing, a key advantage of the ultrasonic tool is that it provides more detailed information about the entire perimeter of the well.

9

Laboratory Testing and Evaluation of Well Cements

Cement and cementing materials testing is a complex process that begins at the cement production stage and continues up to the development stage of cement slurry recipes in the service company's laboratories. This process is an integral part of the cementing process.

In general, laboratory testing of cements and cementing materials can be conventionally divided into two main processes:

- Evaluation of the performance characteristics of the cement slurry
- Evaluation of chemical properties of cement slurry

It is the evaluation of performance characteristics of cement slurry that mostly deals with service company labs at the stage of development of cement slurry and performance of cementing jobs. The essence of the process is to study specific properties of a cement slurry or stone under simulated downhole conditions.

Evaluation of chemical characteristics of cement slurry includes quantitative or qualitative analysis of slurry components before and after preparation. The essence of this process is to ensure the proper quality of cement slurry and its components. Such methods are also used to control the quality of the mixing water.

This chapter briefly describes the basic laboratory test procedures and equipment for cement testing. The material presented in this chapter is not intended as a guide or recommendation for laboratory testing. Detailed procedures for testing well cements are given in the relevant industry standards: American Petroleum Institute (API), International Organization for Standardization (ISO), or American Society for Testing Materials (ASTMs).

9.1 Preparation of Cement Slurry

Cement slurry is prepared in the laboratory according to API/ISO standards using a propeller mixer (Figure 9.1). The aforementioned standards strictly prescribe the technical characteristics (i.e. propeller rotation speed, permissible wear of the mixer blades, etc., used for mixing cement slurries) used in the preparation of cementing slurries. As a rule, 600 ml of cement slurry is prepared, this volume is enough for most laboratory tests. Cement is added to mixing water previously poured into a mixer cup for 15 seconds at 4000 rpm, then the speed is increased to 12 000 rpm and continued mixing for 35 seconds.

It is necessary to carefully monitor the condition of the mixer blades since cement is highly abrasive and over time destroys the blades. The wear of the mixer blades is determined by the weight method, i.e. the loss in weight of the blade over time is compared. The permissible amount of weight loss of the blade is prescribed in API/ISO standards and once it is reached, the blade must be replaced with a new one.

Dry materials are mixed with cement powder until a homogeneous mixture is obtained before adding it to the mixing fluid. Liquid additives are mixed with water and the resulting mortar is called the mixing fluid. It should be noted that the order of adding liquid additives to water is a very critical factor and in some cases, incorrect sequence of additive mixing can significantly worsen physical and chemical properties of cement slurry. It is a good practice to document the order and time of additives mixing when preparing cement slurry.

Figure 9.1 Cement slurry mixers (Fann Instrument Company).

If cementing will involve batch (cyclic) mixing of cement slurry, i.e. mixing and dispensing finished slurry in batches, then after preparation, the cement slurry should be transferred to a consistometer and conditioned according to expected operating conditions (i.e. time and temperature).

However, standard mixing procedures are not acceptable for cement slurries that have been made lightweight with the addition of microspheres or nitrogen. Hollow microspheres easily break down at high paddle speeds. In general terms, the mixing procedure is as follows: cement with added microspheres is added to the mixing water at a maximum speed of 4000 rpm for 30 seconds and at this speed, the mixing continues for another 300 seconds. Foamed cement mortars are often prepared in special mixers at higher mixing speeds. The disadvantage of this procedure is that it poorly simulates downhole conditions, particularly reservoir pressure.

9.2 Test Methods of Cement Slurries

9.2.1 Density

Mud balances are used to determine the density of cement slurry (Figure 9.2). The procedure is as follows:

- Cement slurry is poured into a measuring cup of mud balance and a cup lid is screwed on tightly.
- The plunger is connected to the lid and is pre-filled with solution.

Figure 9.2 Pressurized mud balance for determining the density of cement slurry (Fann Instrument Company).

- The slurry being in the piston is squeezed out into the cup until it is impossible to squeeze additional cement slurry inside. In fact, it is required to pump the cement slurry under pressure in order to make sure that there are no air bubbles in the slurry and it is not required to apply titanic efforts while pumping the cement into the cup.
- The instrument is then placed on a fulcrum and a state of equilibrium is established by adjusting the position of the sliding weight, that is, until the instrument is tilted to one side or the other and is parallel to the laboratory table or other surface on which the measurement is made. The arm over which the weight slides is graduated and the position of the sliding weight at which the instrument is balanced will indicate the density of the solution being measured.

9.2.2 Thickening Time

Cement slurry thickening time is a parameter indicating the length of time that the cement slurry remains in a pumpable, fluid state under simulated downhole conditions. Tests are conducted on a consistometer, a device that measures consistency of cement slurry in a consistometer cup rotating at a constant speed under simulated downhole conditions.

There are two basic types of consistometers:

- High-pressure, high-temperature (HPHT) consistometers
- Atmospheric consistometers

HPHT consistometers are standardized to simulate downhole temperatures up to 204 °C and pressures up to 175 MPa, but there are models with maximum possible temperatures and pressures of 371 °C and 280 MPa, respectively (Figure 9.3).

Atmospheric consistometers are designed to simulate low-temperature well conditions (Figure 9.4). An important feature is that they are also tested without pressure (hence the name). Today, however, these instruments are more commonly used to prepare

Figure 9.3 High pressure and high temperature consistometer (Fann Instrument Company).

cement slurries (according to various API/ISO procedures) prior to performing rheological, fluid loss and free fluid than they are used to measure consistency.

As noted earlier, consistometers determine the dynamics of cement slurry consistency under downhole conditions; this parameter is measured in Bearden units (Bc), a dimensionless value with no direct conversion factor to more common viscosity units, such as Pa-s or poise. When the mortar reaches a consistency of 100 Bc, the experiment is considered over, but in practice, most service companies accept a consistency of 70 Bc as the maximum pumpable consistency of the cement slurry.

Figure 9.4 Atmospheric consistometer (Fann Instrument Company).

Figure 9.5 shows the result of a typical thickening-time test. The consistency curve often begins with a flat segment reflecting the low consistency values of the slurry, which can last up to several hours. As the solution thickens, the curve begins to rise to a maximum of 100 Bc. The point in the curve where the thickening curve begins to rise is called the point of departure (POD). The consistometer cup, in which cement slurry is poured, is a prefabricated metal construction, which does not provide any ability to simulate fluid loss, which, in turn, may result in different slurry thickening times in the laboratory and in actual downhole conditions. This difference will be especially significant in the case if cement slurry does not contain additives regulating fluid loss. Well temperature and pressure also influence the time of cement slurry thickening. It is important to simulate the rate of pressure and temperature increase as close as possible to the planned cementing job during the tests. The time it takes the cement slurry to reach a given temperature (downhole circulation temperature) and pressure (downhole pressure) during a test is determined by the well design and the injection rate of the cement slurry. In essence, the purpose of performing this test is to ensure that the slurry formulation is correct and will provide the required time for cementing operations.

For many years, API standards contained a set of thickening time curves derived from field data that indicated rates of temperature and pressure rise and their final test values. Today those curves have been replaced by mathematical equations that enable calculations of all required thickening time test parameters for a specific cementing job using specialized software.

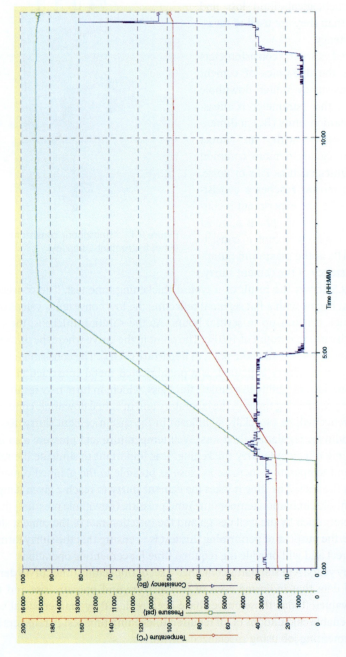

Figure 9.5 An example of the result of measuring the consistency of a cement slurry.

An alternative method of compiling test procedures is based on using calculated bottom hole circulation temperature (BHCT) data for wells with different depths and geothermal gradients. These data are tabulated in API standards. In brief, the procedure for selecting this method is as follows:

1) If cement slurry will be mixed batchwise (cyclically) during cementing, then the slurry is mixed in the consistometer cup at the predicted temperature at the wellhead for an identical period of time. This cementing step is simulated prior to the start of the test for the thickening time.
2) Time of displacement of the first batch of cementing slurry to bottom hole by the formula is calculated:

$$t_{disp} = \frac{V_c}{q}$$

where q – speed of cement slurry pumping (m³/min), t_{disp} – time of displacing the first pack of cement slurry downhole (min), and V_c – casing volume (m³).

3) Downhole pressure is calculated by the formula:

$$P_{bh} = g \times \rho_{dm} \times D_{TV}$$

where D_{TV} is true vertical depth in the upper part of the cement column (m), g is acceleration of gravity (m/s²), P_{bh} is bottom hole pressure (kPa), and ρ_{dm} is drilling mud density (kg/m³).

4) Initial cementing pressure (P_{in}), i.e. pressure, to which the first pack of cementing slurry is exposed upon leaving the cementing head, has to be calculated.
5) The rate of pressure increase as the cement slurry advances to the bottom hole is calculated using the formula:

$$R_{pres} = \frac{P_{bh} - P_{in}}{t_{disp}}$$

where R – pressure increase rate, P – bottom hole pressure (kPa), P – initial pressure (kPa), and t – time of displacing the first pack of cement slurry to the bottom hole (min).

6) Bottom hole circulation temperature should be determined according to API RP 10B tables.
7) Temperature increase rate is calculated by subtracting ambient temperature from bottom hole circulation temperature and dividing by the time required for cement slurry pumping downhole.

It should be noted that when testing to determine the time of cement slurry thickening, regardless of the method of calculating the required parameters, one important assumption is made, that the cement slurry reaches the maximum values of temperature and pressure acting on it simultaneously. In practice, this is not always true, since, for example, a bundle of cement at the top of a cement sheath is almost certainly exposed to higher temperatures and pressures as it circulates through the deeper intervals of the well. If such data is available, it is advisable to build the testing procedure according to it.

9.2.3 Fluid Loss

Fluid loss is a parameter that shows the ability of cement slurry to release water into porous rocks under differential pressure during cementing operations and immediately after the completion of cementing operations. The testing procedure is as follows:

- The prepared slurry is conditioned on an atmospheric consistometer at circulating temperature at the bottom hole.
- The solution is placed in a preheated unit cell for measuring fluid loss and exposed to the 6.9 MPa differential pressure. The differential pressure causes the fluid to filter through a standard 325 mesh (45 μm) filtration screen attached to a 60 mesh (250 μm) screen. The filtration area is 22.6 cm².
- After 30 minutes, the volume of the released filtrate is measured. The obtained value is multiplied by two and is the fluid loss rate of the given solution. If the fluid is released for less than 30 minutes, the following equation is used to calculate the API fluid loss rate (q_{API}):

$$\left(q_{API}\right)_{calc.} = 2V_f \left(\frac{5477}{\sqrt{t}}\right)$$

where V_f – volume of filtrate (ml) collected during the time interval t (min).

The tests are performed either with a static filter press or a dynamic filter press (Figure 9.6).

There is a slight difference in the testing procedure depending on the type of filter press used. If a static filter press is used, the cement slurry is conditioned prior to testing and then transferred to a filter press cell preheated to test temperature. When a dynamic filter press is used, conditioning takes place directly in the instrument itself with constant agitation. Regardless of the instrument used, however, the solution is actually always in a static state when the test itself is performed.

It should be noted that the procedure described here for measuring cement slurry fluid loss simulates the time interval after cement injection has stopped, that is, when cement has already been injected into the target intervals, and does not in any way simulate cement slurry fluid loss during the injection process itself.

Figure 9.6 Filter presses for cement slurry fluid loss tests (Fann Instrument Company).

9.2.4 Free Water

After cementing is completed, there may be water separation in the static cement slurry. The released water migrates upward and accumulates in the upper part of the cement sheath, which leads to failure of insulation tightness, especially in inclined wells. Under laboratory conditions, this feature of cement slurry is tested using a graduated cylinder of simulating wellbore (Figure 9.7). The test duration is two hours from the time the slurry is poured into the graduated cylinder. At test

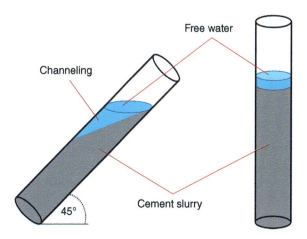

Figure 9.7 Free water.

temperatures less than 80 °C, the graduated tube is placed in a preheated autoclave. At higher temperatures, the graduated tube is placed in a preheated autoclave filled with oil where pressure is maintained high enough to prevent the solution from boiling. To simulate an inclined wellbore, in many laboratories, the graduated cylinder is placed at an angle of inclination of the well, as a rule, the obtained values of water separation in this case increase, but there is no generally accepted explanation of the mechanism of this phenomenon.

9.2.5 Sedimentation Test

When the cement slurry is in a static position, in addition to free water, separation and sedimentation of suspended solids in the slurry are possible, which may lead to changes in slurry density and, as a consequence, penetration of cement slurry into the formation. Probability of such an effect is considerably higher in weighted cement slurries. Laboratory procedure for determination of this parameter consists in keeping cement slurry under certain conditions in a tube of special design. The slurry is preconditioned under borehole conditions and then poured into a test tube, which, in turn, is placed in a water bath or autoclave preheated to the expected borehole temperature or 90 °C (whichever is lower). During the test, the measurement temperature is adjusted according to the temperature conditions in the wellbore. It is common practice to build up some pressure in the autoclave to prevent the slurry from boiling. After an appropriate curing period (usually 24 hours), a cylindrical-shaped cement stone specimen is taken from the tube and cut into several equal segments. The density of each segment is measured and the density difference between the cement slurry and stone is calculated according to the formula:

$$\Delta\rho = \left(\frac{\rho_{\text{cement stone}} - \rho_{\text{slurry}}}{\rho_{\text{slurry}}} \right) \times 100\%$$

where $\rho_{\text{cement stone}}$ is the density of cement stone, ρ_{slurry} is the density of cement slurry, and $\Delta\rho$ is the difference in the densities of cement slurry and stone expressed as a percentage.

9.2.6 Rheological Measurements

Rheological properties of cement slurry are one of the most important parameters, on which the success of cementing operations depends. However, before describing basic measurement methods, it is necessary to briefly outline basic rheological concepts and terms.

9.2.6.1 Flow Types

Under steady-state and isothermal conditions, fluids flow in either laminar or turbulent flow modes. These flow modes are separated by a transition zone.

For a more simplified explanation of these concepts, we will further consider the flow of fluid in a pipe.

9.2.6.2 Laminar Flow

At the laminar flow of liquid in the pipe, separate liquid particles move along the flow direction along trajectories corresponding to straight lines situated parallel to the pipe axis (Figure 9.8). The velocity of the particles decreases from the center to the pipe wall and for most liquids, the velocity of the particles in contact with the pipe wall is practically zero. The shape of the velocity profile varies from liquid to liquid depending on rheological properties. It is worth noting that for some liquids, the particle velocity is practically uniform in the central part of the pipe and sharply decreases to zero near the pipe wall (Figure 9.9). Such flow is called plug flow.

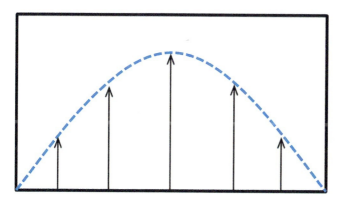

Figure 9.8 Flow velocity profile for Newtonian fluid in laminar flow in a pipe.

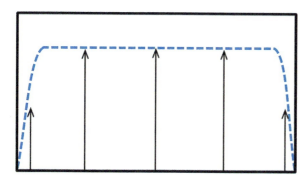

Figure 9.9 Flow velocity profile for Newtonian fluid in plug fluid flow.

9.2.6.3 Turbulent Flow

In turbulent fluid flow, the movement of particles has a vortex character and is quite different from the smooth movement in laminar flow (Figure 9.10). The fluid particles do not move parallel to the pipe walls, but chaotically and their velocity depends on time. In this mode of fluid flow, there is a constant transfer of momentum from one region to another. In turbulent flow, the flow velocity increases rapidly away from the walls of the pipe and becomes almost constant in the central part of the pipe.

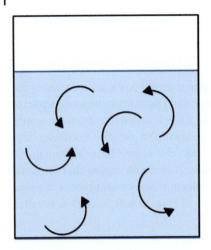

Figure 9.10 Turbulent flow.

At low flow velocities, liquids always flow in laminar mode, and their behavior can be characterized by a property called viscosity, which will be discussed later. As flow velocity increases until a turbulent regime is established, there is a transition mode, and the flow becomes less dependent on viscosity and more dependent on inertial forces.

9.2.6.4 Basic Rheological Concepts

In the laminar flow regime, fluid motion can be compared to a large number of plates moving parallel to each other at different speeds (Figure 9.11).

In this simple flow geometry, the velocity of the fluid particles changes linearly from one plate to the next. The shear rate (or velocity gradient) is mathematically described by the equation:

$$\text{Shear rate} = \frac{\text{Velocity difference between the two plates}}{\text{Distance between two plates}}$$

or

$$\frac{dv}{dx} = \frac{v_1 - v_2}{L} \tag{9.1}$$

where x – axis is perpendicular to the plates.

The dimensions of Eq. (9.1) are as follows:

$$\frac{\text{distance} \times \text{time}^{-1}}{\text{distance}} = \text{time}^{-1}$$

Thus, the unit of shear rate ($'Y'$) is s^{-1}.

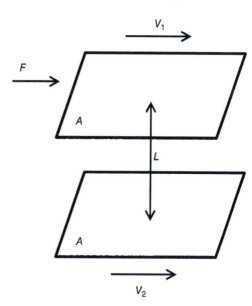

Figure 9.11 Fluid flow between two parallel plates.

The shear stress, denoted by τ, is the force per unit surface area that causes shear

$$\tau = \frac{F}{A} \tag{9.2}$$

In common oilfield units, the unit of measure for shear stress is pound-force/100 ft². In the SI system, this unit is the pascal (Pa). Fluid viscosity is the ratio of shear stress (τ) to shear rate (γ), referred to as μ.

$$\mu = \frac{\tau}{\gamma}$$

In common oilfield units, the unit of viscosity is centipoise (cP). In the SI system, this unit is pascal-second (Pa-s).

Returning to the simple case of laminar flow in a pipe, we can assume that shear stress is proportional to hydraulic friction (or frictional pressure loss). For the sake of simplicity, we will assume that the shear rate is proportional to the flow velocity, but this statement is not scientifically correct. So the viscosity of the fluid is a parameter characterizing the dependence of hydraulic friction on the flow velocity. It depends on temperature, pressure, and, in some cases, on shear rate.

9.2.6.5 Rheological Models

Liquids are classified as Newtonian and non-Newtonian according to the relationship between shear stress and shear rate in steady laminar flow.

9.2.6.6 Newtonian Fluids

Newtonian fluids conform to the Newtonian model in which the shear stress (τ) is directly proportional to the shear rate (γ) and is described by the equation:

$$\tau = \mu \times \gamma$$

This relationship is illustrated in Figure 9.12.

The slope of the line is proportional to the viscosity (μ) of the liquid, which is a constant quantity that does not depend on the flow conditions, but only on temperature and pressure. Common Newtonian fluids include water, gasoline, etc.

9.2.6.7 Non-Newtonian Fluids

Non-Newtonian fluids describe fluids in which the rheological behavior cannot be described by the classical Newtonian model (i.e. the relationship between shear stress and shear rate differs from a straight line through a point of reference). In addition to being a function of temperature and pressure, the viscosity of these fluids can either decrease with shear rate (in which case it is called pseudoplastic) or increase with shear rate (in which case it is called dilatant). Most drilling and cement slurries are pseudoplastic fluids. To describe the rheological properties of fluids used in well cementing, the most commonly used models are:

- Power-law model
- Bingham model
- Herschel–Bulkley model

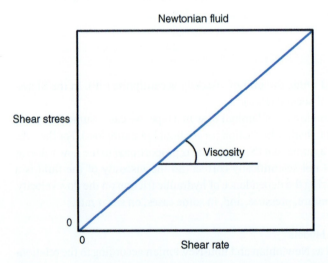

Figure 9.12 Dependence between shear rate and shear stress for Newtonian fluid.

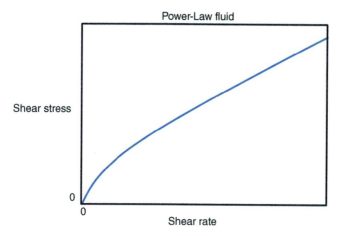

Figure 9.13 Dependence between shear rate and shear stress for Power-law fluid at $n < 1$.

9.2.6.8 Power-Law Model

Fluids subjected to this model (Figure 9.13) are described by the following equations:

$$\tau = k\gamma^n$$
$$\mu = k\gamma^{n-1}$$

where k is the flow consistency index (in the SI unit of measure is Pa*s^n), and n is the index of fluid behavior (dimensionless).

Fluids whose behavior is described by a power law can be subdivided into three different types of fluids, depending on their behavior index:

$n < 1$ Pseudoplastic
$n = 1$ Newtonian fluids
$n > 1$ Dilatant fluids

A fluid is described by a power law only in laminar flow, the shear stress increases much faster when it switches to turbulent mode.

9.2.6.9 The Bingham Model

This model describes liquids having an initial yield strength τ_0, below which they do not flow and have solid properties (Figure 9.14). The rheological behavior of such fluids obeys Newton's law when $\tau > \tau_0$.

Two parameters define the Bingham plastic model:

- the value of τ for $\gamma = 0$, i.e. τ_0 yield stress (initial shear stress)
- slope of the straight line, μ_p – plastic viscosity

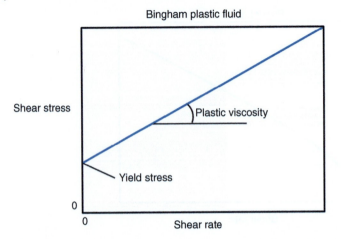

Figure 9.14 Dependence between shear rate and shear stress for the Bingham fluid.

Fluids whose rheological behavior obeys this model are described by the following equations:

$$\tau = \tau_0 + \mu_p \gamma, \text{при } \tau > \tau_0$$

$$'Y = 0, \text{при } \tau \leq \tau_0$$

$$\mu = \mu_p + \frac{\tau_0}{\gamma}$$

At transition of a flow into turbulent mode, as well as in other mathematical models describing rheological behavior of a liquid, the shear stress grows much faster than it is described in the model.

9.2.6.10 Herschel–Bulkley Model

This model describes fluids whose rheological behavior combines the behavior of fluids described by the power and Bingham models (Figure 9.15). These fluids have a yield stress, and above this value, the dependence of shear rate on shear stress follows a power law. Rheological behavior of these fluids is described by the following equations:

$$\tau = \tau_0 + k\gamma^n \text{ at } \tau > \tau_0$$

$$\mu = \frac{\tau_0 + k\gamma^n}{\gamma}$$

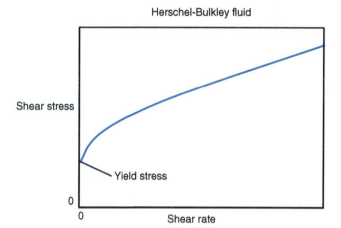

Figure 9.15 The relationship between shear rate and shear stress for a fluid described by the Herschel–Bulkley Model.

At transition of flow into turbulent mode, as well as in other mathematical models describing rheological behavior of fluids, shear stress grows much faster than described in the model.

Two more mathematical models should be noted, which are most commonly used in evaluating rheological properties of fluids used in cementing.

The Casson model, described by the equation:

$$\tau = \left(\sqrt{\tau_0} + \sqrt{\mu_p \times \gamma}\right)^2$$

Robertson and Stith's model, described by the equation:

$$\tau = \left[(\tau_0)^{\frac{1}{n}} + k^{\frac{1}{n}}\gamma\right]^n$$

Rheological measurements of cement slurries are carried out on rotary viscometers, which are a construction of two rotating bodies, aligned on the axes (Figure 9.16). The space between the bodies is filled with investigated solution.

Figure 9.16 Rotary viscometer (Fann Instrument Company).

Table 9.1 Example of the results of rheological measurements on a rotary viscometer.

Speed of rotation, rpm	Viscometer readings with increasing rotation speed (ramp up)	Viscometer readings at decreasing rotation speed (ramp down)	Ratio of obtained viscometer readings	Average values of viscometer readings
3	21	24	0.87	22.5
6	40	36	1.11	38
30	65	83	0.78	74
60	74	100	0.84	92
100	100	115	0.87	107.5
200	137	147	0.93	142
300	170			170

In this case, one of the bodies is rotated, while the other remains stationary. The investigated solution transmits the rotation from the moving body to the static body. The speed at which the rotation is transferred from one body to the other determines the viscosity of the substance. After the solution has been prepared, it is conditioned, but the temperature in the consistometer cup should not exceed room temperature and the solution should be heated to test temperature directly in the consistometer with constant agitation for 20 minutes.

If preconditioning was performed in the consistometer under increased temperature and pressure, the solution should be cooled down as quickly as possible to 88 °C before the slurry container is opened.

Then the solution is transferred to the cup of the viscometer and measurements are made. The readings are taken first in ascending order (ramp up) and then in descending order (ramp down), the results are presented as the average of these readings (Table 9.1).

9.2.7 Static Gel Strength (SGS)

There are three basic ways to determine the static gel strength (SGS) of cement slurry: using a rotary viscometer, the ultrasonic method, and instruments that allow the determination of SGS by constant or pulsed rotation. Today, the ultrasonic method is excluded from the recommended API test methods.

When measuring SGS with a rotary viscometer, measurements are usually made immediately after determining the rheological properties of the solution. The gel strength is measured at two points after the solution has been left in a static position for 10 seconds and 10 minutes (API RP 10B). This measurement method is fairly straightforward but does not accurately simulate downhole conditions.

Figure 9.17 MACS II multipurpose cement slurry analysis system (Fann Instrument Company).

There is only one device in the market today that allows making SGS measurements in as close as possible to downhole conditions, MACS II from FANN Instruments (Figure 9.17). The device, in fact, is a multitasking cement slurry analysis system and determines SNA by rotation at very low velocities. This device, which acts as an atmospheric consistometer as well as a high-pressure consistometer, allows the analysis of the thickening time and SGS of the cement mortar.

9.2.8 Flowability of Cement Slurries

When testing cement slurry according to GOST, one of the most widely used tests to investigate the pumpability of cement slurry is to determine the flowability of cement slurry. This parameter is determined with the help of a device developed by Azerbaijan Scientific Research Petroleum Institute (AzNII), called "AzNII cone" (Figure 9.18). It consists of a truncated cone with an inner diameter of the upper base of 36 ± 0.5 mm and the lower of 64 ± 0.5 mm and a volume of $120 \, cm^3$. The cone is placed on the glass, under which a round crossed by concentric circles at intervals of 5 mm is placed. With the help of adjusting screws, serving at the same time as instrument supports, the circle with the glass is set horizontally on the level. The cone must be set in the center of the circle. To prevent premature

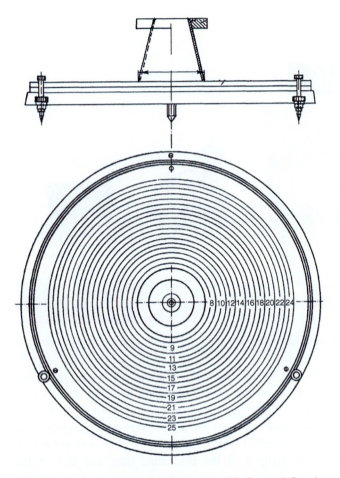

Figure 9.18 Cone of AzNII (Azerbaijan Scientific Research Petroleum Institute) to determine the flowability of cement slurry.

lifting of the cone when filling it with cement mortar, the cone is weighted to 300 g by soldering a ring to its outer surface or by thickening the walls of the cone. The inner surface of the cone must be polished and free from irregularities or roughness. For the test, a cement slurry of 250 cm^3 shall be prepared. The prepared cement slurry is poured into the cone up to the level of the upper ring and the cone is smoothly lifted vertically upwards. Cement mortar spreads on the glass. Using concentric circles in mutually perpendicular directions determines the largest and smallest diameters of the circle of dissolution. According to them, the average diameter is calculated, which characterizes the flowability of the solution.

To get correct results, the inner surface of cone and glass should be clean and dry. The flowability of the slurry is expressed in centimeters.

Value of flowability ensuring normal pumpability of cement slurry should not be less than 18 cm.

9.3 Test Methods of Cement Stone

In the wellbore, the cement sheath supports the casing and seals the annular space between the casing and the formation or between two casing strings. Over the life of the well, the cement stone is subjected to various stresses caused by changes in temperature and pressure. Uniaxial compressive strength and gas or water permeability are the main parameters determined for this purpose. However, they do not provide sufficient information to predict and simulate the behavior of the cement sheath under the expected stress conditions in the well. Analytical and numerical models help solve this problem by predicting the distribution of stresses in the cement sheath caused by changes in downhole conditions. Regardless of the calculation method chosen, the input data are the mechanical and thermal properties of all materials involved-the steel casing, the cement and the formation. Results can vary significantly depending on the test method used to determine these properties. In order to obtain consistent, reproducible, and comparable results, test procedures must be strictly adhered to.

9.3.1 Mechanical Strength of Cement

Laboratory tests for determining the strength of cement stone are prescribed in API standards, according to which there are two ways of measuring strength:

- The destructive test that determines the compressive strength values of cement stone by a destructive compression test
- The non-destructive test, which is based on sound velocity changes as sound is transmitted through a cement slurry specimen during the curing process

9.3.2 Destructive Test (Compressive Strength)

The cement slurry for the test is prepared according to API/ISO procedures, poured into 2-in. cube molds (Figure 9.19), and cured for various periods of time at specific temperatures and pressures. The resulting cement stone cubes are removed from the molds and placed in a hydraulic press, where an increasing single-axis load is applied to each cube until it fractures. The compressive strength is then calculated by dividing the force at which failure occurred by the cross-sectional area of the specimen.

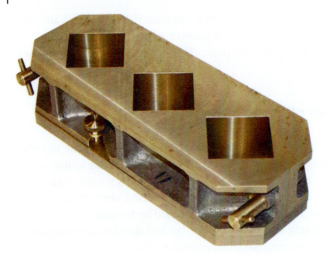

Figure 9.19 Cubic molds for cement stone strength tests (Fann Instrument Company).

9.3.2.1 Non-destructive Test (Ultrasonic Measurement)

The Ultrasound Cement Analyzer (UCA), shown in Figure 9.20, measures the transit time of the ultrasonic energy signal through a cement sample as it cures under simulated temperature and pressure conditions. The ultrasonic measurement is non-destructive and can be performed continuously as the cement sample cures at high pressure and elevated temperature (Figure 9.21).

9.3.3 Expansion and Shrinkage

Cement expansion during setting is measured according to the procedure described in ASTM C151. The essence of the test is to place the cement slurry in a special bar-shaped mold for curing under water at atmospheric pressure. After the cement bar has attained sufficient strength, it is removed from the mold and its length is measured, and then the bar is placed back in the water bath for further curing. Periodically, over a period of time, the bar is removed for additional length measurements. Changes in the length of the bar are taken as indications of cement mortar expansion during curing. However, this method has two major drawbacks:

- Because the cement must gain a certain strength before measuring the length, it is not possible to obtain reliable initial values
- This method of measurement does not take pressure into account.

In July 1997, the API published a report describing equipment and procedures for investigating shrinkage and expansion in oil well cements.

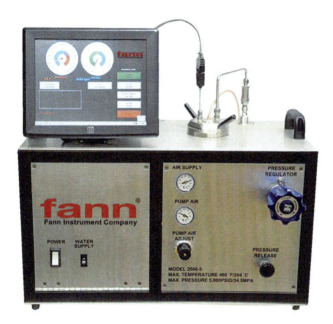

Figure 9.20 Ultrasonic cement analyzer (Fann Instrument Company).

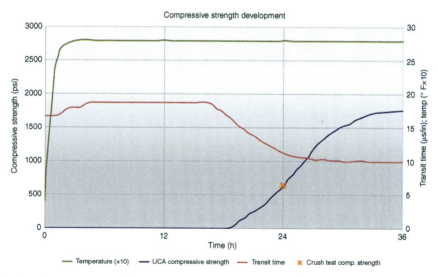

Figure 9.21 Example of a result of measuring the compressive strength of cement with an ultrasonic cement analyzer.

9.3.4 Gas Migration

To date, neither API nor ISO has published a standard method or procedure for testing gas migration control in cement slurries; that is, there is no generally recognized standard laboratory procedure for determining the ability of a cement system to prevent or reduce gas migration.

Over the years, a significant number of nonstandard laboratory tests have been developed to determine this parameter. In general, they can be summarized in two main approaches: creating a large-scale experimental setup to simulate processes occurring in the wellbore or studying individual parameters on smaller-scale models. Not uncommon is the use of various mathematical models, the input data for which are such parameters as static shear stress and shrinkage of cement stone.

9.3.5 Cement Stone Permeability

The permeability of the cement sheath plays an important role in ensuring the tightness of the insulation. The procedure for determining the relative permeability of a cement sheath over water or gas is specified in API standards.

The permeability to water is determined by squeezing water through a cement stone sample at a pressure difference of 100–1400 kPa. The filtrate is collected for 15 minutes or until 1 ml of filtrate has accumulated in the measuring beaker. The Darcy equation is used to calculate permeability.

To determine gas permeability, a cement stone sample should be pre-dried either in a special drying oven or in a high-temperature roller oven. API and ISO standards do not regulate this stage of sample preparation and in fact each laboratory follows its own developed drying procedure. However, this process should be taken very seriously as there is a very high risk of sample failure, which will ultimately lead to questionable measurement results. For the calculation of the results, the Darcy equation is used as well.

9.3.6 Thermophysical Properties of Cement

Changes in wellbore temperature during production can deform the cement sheath and lead to cracking and fractures that compromise the integrity of the cement sheath. The theory of thermoelasticity allows a linear relationship to be established between deformation, stresses and temperature change, assuming that the cement is homogeneous and isotropic. If we consider radial geometry, the temperature distribution as a function of time is determined from the heat diffusion equation, which is expressed with the condition that the initial and boundary conditions depend only on the radius r as:

$$\frac{\partial^2 T}{\partial r^2} + \frac{1}{r}\frac{\partial T}{\partial r} = \frac{\rho C}{\lambda}\frac{\partial T}{\partial t}$$

where λ is thermal conductivity, ρ is density, C is specific heat capacity, and T is temperature.

Due to thermal expansion, temperature affects the stress state of various materials. The linear thermal strain (ε_T) is directly proportional to the temperature change ΔT

$$\varepsilon_T = \alpha \times \Delta T$$

where α is the coefficient of linear thermal expansion.

The test methods for determining the thermal conductivity of cement and the thermal expansion coefficient are described as follows.

9.3.6.1 Thermal Conductivity

Thermal conductivity of cement is determined by the method described in ASTM D2326-70, by placing a probe between two flat surfaces of a test specimen of cement. The sample size must be large enough to cover the probe surface and reduce boundary effects. After thermal stability is reached, a thermal pulse is generated to break the equilibrium. The resulting temperature changes are recorded as a function of time. The linear thermal conductivity of the cement is derived from this relationship. To ensure repeatability, this test must be performed on at least three samples. The higher the thermal conductivity value, the faster the temperature in the cement column will equalize. Modern laboratory equipment automates the testing process and provides a direct reading of the thermal conductivity.

9.3.6.2 Coefficient of Linear Thermal Expansion

The coefficient of thermal expansion is the strain per unit of temperature rise. A simple method of determining this coefficient is by mechanical dilatometry. A typical instrument (Figure 9.22) consists of a quartz rod and a digital micrometer on a base of a thick stainless steel plate.

The stand is in a water bath. The measuring tip of the digital indicator, extended with a thin quartz rod, is in contact with the top of the cement sample. The water bath temperature increases gradually and the maximum temperature does not exceed the curing temperature of the cement. The temperature and deformation of the specimen are recorded throughout the test. The linear coefficient of thermal expansion is calculated from the relative expansion of the specimen and the increase in temperature.

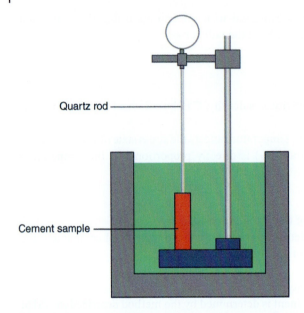

Figure 9.22 Instrument for measuring the linear coefficient of thermal expansion of a cement sample.

9.4 Laboratory Evaluation of Spacers and Washers

Buffer and washer fluids serve two important functions during the cementing operation: cleaning and removing drilling fluid from the wellbore and minimizing cement contamination with drilling fluid. To achieve these goals, both fluids must provide certain degrees of wellbore cleanup and be compatible with both the displaced drilling fluid and the cement slurry being pumped into the well. Laboratory testing of these fluids is all about evaluating these two parameters.

9.4.1 Compatibility of the Buffer/Washer Fluid with the Drilling Fluid and Cement Slurry

The procedure for this test is also described in the API recommendations. The test involves mixing at various volume ratios of buffer fluid, drilling fluid, and cement slurry. Thickening time, fluid loss, and rheological properties of the slurries are measured.

9.4.2 Efficiency of Wellbore Cleaning with Washer Fluid

There are no standard procedures to evaluate this parameter. In fact, the cleaning process takes place in two stages: volumetric displacement of the drilling fluid and subsequent cleaning of the annulus walls. Volumetric displacement is controlled by fluid mechanics, rheology, and density of the buffer or washer fluid. These parameters can be measured under simulated downhole conditions using standard API and ISO procedures as long as the temperature does not exceed 185 °F (85 °C). However, it is often necessary to extrapolate laboratory data to downhole temperature and pressure conditions using appropriate models.

9.5 Chemical Analysis of Mix Water

In the context of well cementing, cement slurry tests prior to cementing operations are often performed using water samples taken directly from the well. Depending on the results, the composition of the cement slurry can be adjusted to achieve the desired characteristics. In cases where complications occur during cementing operations, it is the water used to prepare the cement slurry that is first analyzed. The most significant chemical parameters affecting the characteristics of cement slurries are:

- pH
- ammonia content
- chloride content
- iron content
- organic acids content
- sulfate content

The above parameters can have a negative effect on hydration and setting as a result of interaction with the Portland cement itself or with cement additives. In large laboratories, methods such as atomic absorption spectrometry and inductively coupled plasma spectrometry are commonly used to determine metal content. Organic substances are determined by Fourier Transform Infrared Spectroscopy (FTIR) or chromatographic methods. Electrode testers or portable titration kits are often used in small laboratories or in the field.

10

Typical Calculations for Well Cementing

At present, calculations of parameters required for the planning of cementing works and recipes of cementing compositions, as a rule, are performed with the use of software and there is no necessity to perform them manually. However, in order to understand the relationship and influence of various factors on cementing slurry properties and cementing process as a whole, we will consider the basic types of standard calculations performed for cementing jobs. The following types of standard calculations are presented in this chapter:

- Calculations for preparation of cement slurry
- Calculations for primary cementing
- Calculations applied at the remedial cementing
- Calculations performed in the case of foamed cement applications

All calculations presented in this chapter are in US measurement units since it is common for calculations involving cementing to be performed in US measurement units.

10.1 Slurry Preparation Calculations

API/ISO standards prescribe recommended mixing water concentrations for all major cement types according to API classification and provide data on slurry yields and resulting densities at the specified water-cement ratios (Table 10.1). It should be noted, however, that in practice, it is not possible to adhere to water-cement ratios specified in API/ISO standards when designing cementing slurry, as it is often necessary to adjust water content due to the presence of various additives in the slurry.

Oil and Gas Well Cementing for Engineers, First Edition. Baghir A. Suleimanov,
Elchin F. Veliyev, and Azizagha A. Aliyev.
© 2023 John Wiley & Sons Ltd. Published 2023 by John Wiley & Sons Ltd.

Table 10.1 Main grades of cements according to the API classification.

Cement grade	Mix water (% BWOC)	Density (ppg [kg/m³])	Slurry yield (ft³/sk [m³/t])
A	46	15.6 [1870]	1.18 [0.355]
B	46	15.6 [1870]	1.18 [0.355]
C	56	14.8 [1770]	1.32 [0.397]
D	38	16.45 [1970]	1.05 [0.316]
E	38	16.45 [1970]	1.05 [0.316]
F	38	16.45 [1970]	1.05 [0.316]
G	44	15.8 [1890]	1.15 [0.346]
H	38	16.45 [1970]	1.05 [0.316]

The water content of the cement slurry regulates parameters such as density, viscosity, free water, and cement strength. This parameter is expressed as a percentage by weight of cement dry weight (BWOC).

10.1.1 Specific Gravity of Cement Slurry

The specific gravity of Portland cement varies between 3.10 and 3.25, depending on the raw materials used for its production. In the following calculations, the specific gravity of cement is assumed to be 3.14. In practice, however, the specific gravity of the Portland cement used must be measured for each individual job in order to ensure the accuracy of the calculations.

10.1.2 The Concept of Absolute and Bulk Volumes

The volume of materials used in cementing is characterized by two main parameters: specific and bulk volumes.
The absolute volume is the volume occupied by the material itself without considering the air volume surrounding the particles.
The bulk volume is essentially the sum of the specific volume of the material and the volume of air surrounding it.
As mentioned earlier, Portland cement is packaged in standard 94 lb bags and has a bulk volume of 1 ft³. The absolute volume occupied by such a bag of cement is 3.59 US gallons or 0.48 ft³, assuming a specific density of 3.14.
Specific gravity and volume data are readily available in many technical manuals or from the manufacturer (Table 10.2). However, it is strongly recommended to determine these parameters in the laboratory before formulating the slurries, as there may be some differences in the values depending on the manufacturer.

Table 10.2 Absolute volume and specific gravity of some cementing materials.

Material	Absolute volume (gal/lbm [m³/t])	Specific gravity
Barite	0.0278 [0.231]	4.33
Bentonite	0.0454 [0.377]	2.65
Coal	0.0925 [0.769]	1.30
Gilsonite	0.1123 [0.935]	1.06
Hematite	0.0244 [0.202]	4.95
Ilmenite	0.0270 [0.225]	4.44
Silica sand	0.0454 [0.377]	2.65
NaCl	0.0556 [0.463]	2.15
Fresh water	0.1198 [1.000]	1.00

Absolute volume values for soluble materials are significantly lower than their bulk volume, as they take up almost no additional space. In the case of liquid additives or low concentrations of these additives, the volume of additional space taken up is negligible and may not be taken into account in the calculations. The only exception is the sodium salt additive, as it is added to some solutions in much higher concentrations than other additives, this difference must be taken into account.

10.1.3 Additive Concentration Calculation

Concentrations of most solid additives for cements are expressed as a percentage of dry weight of cement, the same as for mixing water.

For example, expressing that a cement mix contains 40% (BWOC) silica sand means that to each bag of cement weighing £94, £94×0.4 = £37.6 of silica sand is added. The mass of the resulting mixture is, therefore, 94+37.6 = 131.6 lb, and the percentage of silica sand in the mixture:

$$\frac{37.6}{131.6} \times 100 = 28.6\%$$

This specificity of cement slurry formulations must be taken into account when calculating or preparing slurries. In rare cases, the calculation of additive concentrations is carried out by weight of dry mix, but this method is not very practical and has not been widely used. At first sight for an unprepared reader, the accepted way of the calculation of concentrations and units of their measurement cause some bewilderment. In fact, such a system of calculation was developed empirically in order to make it as convenient as possible for the cementing crew. So, for example, barite concentration is calculated in "pound per sack (lbm/sk) of cement," which is much more convenient than making any mathematical calculations. Liquid additive concentrations are usually expressed in gallons per bag of cement (gal/sk).

10.1.4 Density and Yield of the Slurry

Mud density is calculated by adding the masses of the components of the cement slurry and dividing this sum by the sum of the absolute volumes of these components. In other words, to determine slurry density, the total mass of cement, water, and additives must be divided by their total volume. (Eq. (C.1)).

$$\rho_{slurry} = \frac{m_{cement} + m_{water} + m_{additive}}{V_{cement} + V_{water} + V_{additive}} \quad (C.1)$$

where m – weight, V – volume.

Cement slurry yield is the volume of slurry obtained from a unit mass of cement, including all additives and mixing water. Since the mass of cement is often expressed in sacks, cement slurry yield in practice is usually understood as the parameter showing how much cement slurry can be obtained from one sack of cement with all additives and mixing water taken into account. The unit of measure is ft^3/sk. Let us consider an example of calculating the density and yield value of a Class G cement. According to the data given in Table 10.1, the mixing water is 44% by weight of dry cement. For ease of calculation, it is convenient to construct the table later.

1) The mixing water mass is 44% BWOC

$$94\,lbm \times 0.44 = 41.36\,lbm$$

Component	Weight (lbm)	Absolute volume (gal/lbm)	Volume (gal)
Cement	94	0.0382	3.59
Water	41.36	0.1198	4.96
Total	135.36		8.54

$$\rho_{slurry} = \frac{135.36\,lbm}{8.54\,gal} = 15.84\,ppg$$

10.1 Slurry Preparation Calculations

The cement slurry yield (Y) is usually expressed in ft^3/sk; in fact, the slurry yield expressed in gallons has already been found (i.e. 8.54 gal) and it is only necessary to convert this value into ft^3/sk. Knowing that, in $1\,ft^3 = 7.48\,gal$, we find the solution yield expressed in ft^3/sk:

$$Y_{slurry} = \frac{8.54\,gal/sk}{7.48\,gal/ft^3} = 1.14\,ft^3/sk$$

The calculation procedure is exactly the same as for other additives. As an example, let us consider a cement slurry of the following composition:

Components	Concentration
Class G cement	
Silica flour	30%
Fluid loss additive (solid)	1%
Dispersant (liquid)	0.2 gal/sk
Water	44%

1) Calculate the masses of all the components of the mixture
   - Mix water (BWOC)

     $94\,lbm \times 0.44 = 41.36\,lbm$

   - Silica flour (BWOC)

     $94\,lbm \times 0.3 = 28.2\,lbm$

   - Fluid loss additive (BWOC)

     $94\,lbm \times 0.01 = 0.94\,lbm$

   - Liquid dispersant

     $$\frac{volume\,(gal)}{absolute\,volume\,(gal/lbm)} = \frac{0.2\,gal}{0.1014\,gal/lbm} = 1.97\,lbm$$

2) Put the results into a table
   Note: Specific volume values are taken from technical handbooks.

Components	Weight (lbm)	Absolute volume (gal/lbm)	Volume (gal)
Cement	94	0.0382	3.59
Water	41.36	0.1198	4.96
Silica flour	28.2	0.0454	1.28
Fluid loss additive	0.94	0.0932	0.088
Dispersant	1.97	0.1014	0.2
Total	166.47		10.12

The volume of silica flour is determined by multiplying the mass and absolute volume:

$$28.2 \, \text{lbm} \times 0.0454 \, \frac{\text{gal}}{\text{lbm}} = 1.28 \, \text{gal}$$

3) Determine the yield and density of the slurry:

$$\text{Density} - \rho_{\text{slurry}} = \frac{166.47 \, \text{lbm}}{10.12 \, \text{gal}} = 16.45 \, \text{ppg}$$

$$\text{Yield} - Y_{\text{slurry}} = \frac{10.12 \, \text{gal/sk}}{7.48 \, \text{gal/ft}^3} = 1.35 \, \text{ft}^3 / \text{sk}$$

10.1.5 Special Additives

10.1.5.1 Sodium Salt

Sodium salts (NaCl) are, due to their availability, quite common additives for cement slurries. The concentration of NaCl is usually expressed as a percentage by weight of water (BWOW). As this additive is used in relatively high concentrations and its specific gravity varies with concentration, this fact must be taken into account in the calculations (Table 10.3).

As an example, let us calculate the density and yield of the following cement slurry:

Components	Concentration
G class cement	
Sodium salt (NaCl)	30%
Water	44%

Table 10.3 Absolute volume of NaCl-a aqueous solution at 80 °F (26.7 °C).

Concentration NaCl (% BWOW)	Absolute volume (gal/lbm [m³/t])
2	0.0371 [0.310]
4	0.0378 [0.316]
6	0.0384 [0.321]
8	0.0390 [0.326]
10	0.0394 [0.329]
12	0.0399 [0.333]
14	0.0403 [0.336]
16	0.0407 [0.340]
18	0.0412 [0.344]
20	0.0416 [0.347]
22	0.0420 [0.351]
24	0.0424 [0.354]
26	0.0428 [0.357]
28	0.0430 [0.359]
30	0.0433 [0.361]
34	0.0439 [0.366]
37.2	0.0442 [0.369]

1) Calculate the masses of all the components of the mixture
   - Mix Water (BWOC)

     $94 \, lbm \times 0.44 = 41.36 \, lbm$

   - Mass of sodium salt (NaCl) (BWOW)

     $41.36 \, lbm \, (\text{water}) \times 0.3 = 12.4 \, lbm$

2) Determine the specific volume of sodium salt at this concentration. According to Table 10.1, it is 0.0433 gal/ft.
3) Put the results into a table

Components	Weight (lbm)	Absolute volume (gal/lbm)	Volume (gal)
Cement	94	0.0382	3.59
Water	41.36	0.1198	4.96
Sodium salt	12.4	0.0433	0.54
Total	147.76		9.09

4) Determine the yield and density of the slurry

$$\text{Density} - \rho_{slurry} = \frac{147.76\,\text{lbm}}{9.09\,\text{gal}} = 16.25\,\text{ppg}$$

$$\text{Yield} - Y_{slurry} = \frac{9.09\,\text{gal/sk}}{7.48\,\text{gal/ft}^3} = 1.21\,\text{ft}^3/\text{sk}$$

10.1.5.2 Fly Ash

As mentioned earlier, fly ash is a pozzolanic extender additive. The composition of a cement mixture containing fly ash is usually written in terms of the specific volume ratio of fly ash and cement. For example, 30 : 70 means that the mixture contains 30% fly ash and 70% cement, with the first number always denoting fly ash and the second number cement. The mass of fly ash required to make 3.59 gallons (absolute volume) of cement mixture is calculated using the following formula:

$$m_{flyash} = m_{repcem}\left(\frac{\gamma_{FA}}{\gamma_{Cement}}\right)$$

where m_{repcem} – mass of cement replaced by fly ash, γ_{Cement} – specific gravity of cement, γ_{FA} – specific gravity of fly ash.

The mass of replaced cement is determined by subtracting the mass of the remaining cement from 94 lbs. For a cement mix of 30 : 70, the mass of cement replaced is calculated as follows:

For the calculation, assume specific gravity of cement and fly ash equal to 3.14 and 2.46, respectively.

1) Calculate the mass of the replaced cement:

$$m_{repcem} = 94\,\text{lbm} - (0.7 \times 94\,\text{lbm}) = 28.2\,\text{lbm}$$

2) Calculate the mass of fly ash:

$$m_{flyash} = 28.2 \times \frac{2.46}{3.14} = 22.09\,\text{lbm}$$

3) Calculate the mass of cement in the mixture:

$$m_{cement} = 0.7 \times 94\,\text{lbm} = 65.8\,\text{lbm}$$

It should also be noted that the mass of the cement mix (i.e. fly ash + cement) with a volume equal to that of a 94-pound cement sack (i.e. 3.59 gal.) is called the equivalent cement sack mass. In the previous example, it would be:

$$\textbf{Equivalent sack}\left(\text{eq sk}\right) = 22.09 + 65.8 = 87.89 \, \text{lbm}$$

For such cement mixtures, the weight of an equivalent cement sack, rather than the weight of a standard cement sack, is used in calculating the concentration. For example, if a reagent of 1% dry weight equivalent (BWOC) were to be added to the above formulation, this would mean 1% of an 87.89 lbm equivalent sack.

Also pay attention to the fact that the physicochemical properties of the fly ash may vary slightly from one supplier to another and the specific gravity has to be measured for each batch separately. Such heterogeneity in physical and chemical properties of the material certainly affects the amount of water needed for slurry mixing.

For example, we calculate the density and yield of a cement slurry of the following composition:

50 : 50 fly ash/cement class G mix, 3% bentonite, and 54% water. The specific gravity of fly ash is taken as 2.48 and absolute volume as 0.0483 gal/lbm.

1) Calculate the mass of replaced cement:

$$m_{\text{repcem}} = 94 \, \text{lbm} - \left(0.5 \times 94 \, \text{lbm}\right) = 47 \, \text{lbm}$$

2) Calculate the mass of fly ash:

$$m_{\text{fly ash}} = 47 \times \frac{2.48}{3.14} = 37.12 \, \text{lbm}$$

3) Calculate the mass of cement in the mixture:

$$m_{\text{cement}} = 0.5 \times 94 \, \text{lbm} = 47 \, \text{lbm}$$

4) Calculate the mass of an equivalent cement sack:

$$m_{\text{eq.sk.}} = 37.12 + 47 = 84.12 \, \text{lbm}$$

5) Calculate the mass of mixing water:

$$m_{\text{water}} = 0.54 \times 84.12 = 45.47 \, \text{lbm}$$

6) Put the results into a table:

Components	Weight (lbm)	Absolute volume (gal/lbm)	Volume (gal)
Cement	47	0.0382	1.794
Water	45.47	0.1198	5.44
Fly ash	37.12	0.0483	1.794
Bentonite	2.52	0.0452	0.114
Total	132.11		9.14

7) Determine the yield and density of the slurry

$$\text{Density} - \rho_{slurry} = \frac{132.11 \, \text{lbm}}{9.14 \, \text{gal}} = 14.45 \, \text{ppg}$$

$$\text{Yield} - Y_{slurry} = \frac{9.14 \, \text{gal/sk}}{7.48 \, \text{gal/ft}^3} = 1.22 \, \text{ft}^3/\text{eq.sk.}$$

10.1.5.3 Bentonite

As mentioned earlier, bentonite is a clay mineral added to cement slurry as an extender. The mechanism of action of this additive is based on increasing the amount of mixing water required to prepare the slurry. The addition of bentonite also helps increase the gel strength values. Due to the fact that the degree of bentonite swelling decreases with the increasing salinity of mixing water, in practice, it is prehydrated in fresh water and an aqueous solution of bentonite is added to cement slurry. For complete hydration of bentonite, 30 minutes is enough, which actually does not affect the duration of the cement slurry preparation process. In addition, the efficiency of hydrated bentonite compared to dry powder is 4 to 1, which means that in order to achieve identical fluid parameters corresponding to 20% of nonhydrated bentonite, only a 5% concentration of hydrated bentonite is required. In pre-1989 editions, API Specification 10 recommended adding 5.3% water to each percentage of bentonite (BWOC). However, this is only a general recommendation as the swelling efficiency of bentonite varies from batch to batch and the actual amount of additional water to be added should be based on the desired performance properties of the cement (i.e. fluid loss of cement slurry, gel strength, etc.) as determined by the laboratory.

As an example, let us calculate the density and yield of a cement slurry of the following composition:

Cement slurry, consisting of class H cement, 3% prehydrated bentonite.

10.1 Slurry Preparation Calculations

1) Transfer the data to a table form

Components	%	Weight (lbm)	Absolute volume (gal/lbm)	Volume (gal)	Additional mix water requirement (gal/lbm)	Mix water (gal)
Cement	—	94	0.0382	3.5908	—	4.3
Bentonite	3%	2.82	0.0453	0.1277	0.69	1.9458
Water						
Total						

Calculate the mass of bentonite:

$0.03 \times 94 = 2.82 \, \text{lbm}$

Using table data on the absolute volume of bentonite of 0.0453 gal/lbm, we find its volume:

$2.82 \times 0.0453 = 0.1277 \, \text{gal}$

Calculate the amount of additional mixing water needed, using the appropriate coefficient. For bentonite, it is 0.69 gal/lbm:

$2.82 \times 0.69 = 1.9458 \, \text{gal}$

2) Determine the mass of mixing water and tabulate the results:

Components	%	Weight (lbm)	Absolute volume (gal/lbm)	Volume (gal)	Additional mix water requirement (gal/lbm)	Mix water (gal)
Cement	–	94	0.0382	3.5908	–	4.3
Bentonite	3%	2.82	0.0453	0.1277	0.69	1.9458
Water		52.0483	0.12	6.2458		6.2458
Total		148.8683		9.9643		

Calculate the total volume of water required to prepare this cement slurry. To do this, sum up the volume of water required for mixing cement and bentonite:

$4.3 + 1.9458 = 6.2458 \, \text{gal}$

Knowing the total volume of mixing water, find its mass:

$6.2458 \div 0.12 = 52.0483 \, \text{lbm}$

3) Determine the yield and density of the slurry:

$$\text{Density} - \rho_{\text{slurry}} = \frac{148.8683 \, \text{lbm}}{9.9643 \, \text{gal}} = 14.9402 \, \text{ppg}$$

$$\text{Yield} - Y_{\text{slurry}} = \frac{9.9643 \, \text{gal/sk}}{7.48 \, \text{gal/ft}^3} = 1.33 \, \text{ft}^3/\text{sk}$$

10.1.5.4 Weighting Agents

Cement slurry weighting agents are used to increase slurry density and the calculations involved in the formulation of such a slurry usually involve determining the amount of weighting agent to obtain the required density.

The amount of weighting agent (hematite) required to increase the density of a cement slurry prepared with Class G cement up to 19 lbm/gal is determined.

Note. *The absolute volume of hematite is 0.0242 gal/lb.*

1) Transfer the data to a table form:

Components	Weight (lbm)	Absolute volume (gal/lbm)	Volume (gal)
Cement	94	0.0382	3.59
Hematite	x	0.0242	$0.0242x$
Water	41.36	0.1198	4.96
Total	$(135.36 + x)$		$(8.55 + 0.0242x)$

By solving the equation here, we find the mass of hematite needed:

$$\rho_{\text{slurry}} = 19 \, \text{ppg} = \frac{(135.36 + x) \, \text{lbm}}{(8.55 + 0.0242x) \, \text{gal}}$$

$$x = 50.14 \, \text{lbm}$$

Thus, since the calculation was based on one sack of cement, it takes 50.14 lbm of hematite added per one sack of Class G cement to increase the slurry density to 19 lb/gal or 53% hematite by weight of cement (BWOC).

10.2 Primary Cementing Calculation

The following types of calculations are carried out for primary cementing:

- Volume of cement to fill the annulus. In practice, an additional 10% is usually added to the cement slurry volume values obtained to compensate for cavernousness in the wellbore. However, the amount of additional cement required

may also be different. In this case, this issue is decided on the basis of experience with cementing similar wells in a given field.
- Water content, density, and cement yield. These calculations will be carried out in accordance with the procedures described earlier.
- The pressure required to set the cement plug. This is approximately equal to the difference in hydrostatic pressure between the fluids in the annulus and the casing. For a more accurate reading, friction losses and pump flow rates must be taken into account. On the basis of cementing plug seat pressure information, the type of pump required to achieve the required pressure for the cementing operation will be determined.
- The displacement volume of the cement plug. This is essentially equal to the internal volume of the casing from the wellhead to the stop collar. Often a small excess volume is also pumped to ensure compression of the air in the slurry.
- Hydrostatic pressure on the formation. In order to avoid fracturing during the cementing process, this parameter must be strictly monitored during the entire procedure. The hydrostatic pressure of the fluid column is determined by the formula:

$$P_{hydrostatic} = 0.052 \rho h$$

where h – liquid column height (ft)

$P_{hydrostatic}$ – hydrostatic pressure (psi)
ρ – density (lbm/gal).

If there is more than one fluid in the well, calculations must be made for each fluid separately.

- **The buoyancy force**. During the well cementing process, under certain conditions (i.e. high density of downhole fluids, short casing length, etc.), there is a high probability that the casing will be lifted out of the well by itself. The reason is that the buoyancy created by the fluids in the well exceeds the gravity force created by the casing and the hydrostatic pressure of the fluids inside it.

Let us consider a cementing calculation example for the well shown in Figure 10.1.

Well data
- Surface casing:
  - Diameter: 13 3/8 in.
  - Weight per unit length of the column: 54.50 lbm/ft
  - Depth: 1600 ft
- Open hole:
  - Diameter: 12 1/4 in.
  - Depth: 4950 ft

10 Typical Calculations for Well Cementing

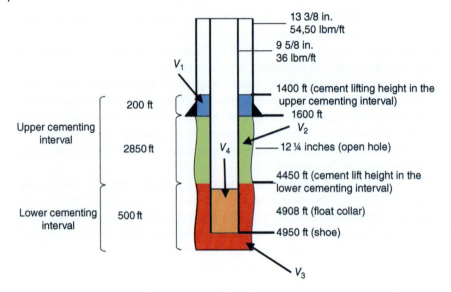

Figure 10.1 Schematic representation of the well design and cementing intervals.

- Drilling mud:
  - Density: 11.50 ppg
- Casing:
  - Diameter: 9-5/8 in.
  - Weight per unit length of the column: 36.00 lbm/ft
- Excessive volume of cement slurry: 25%
- Shoe:
  - Length: 42 ft
- Upper cementing interval:
  - Cement lifting height in the surface casing 13-3/8 in.: 200 ft
  - Cement lifting height in the upper cementing interval: 1400 ft
  - Cement slurry density 13.0 ppg, mortar yield 1.5 ft³/sk
- Lower cementing interval
  - Cement lifting height in the lower cementing interval: 4450 ft
  - Cement slurry density 16.4 ppg, slurry yield 1.05 ft³/sk
- Spacer
  - Density 12.5 lbm/gal
  - Volume 40 bbl
- Displacement fluid:
  - Drilling fluid with density of 11.5 ppg

10.2.1 Volume of Cement Slurry

It should be noted that these calculations use the measured well depth and not the true vertical depth.

1) **The amount of cement required to cement the upper cementing interval (1400–4450 ft).**
   The length of this interval is 3050 ft (i.e. 4450 − 1400 = 3050 ft), 300 ft of which is between the technical and casing strings. The volume of slurry required to cement the upper cementing interval will be the sum of the following annular volumes:
   • The annular volume between the casing and the technical casing:

   $$V_1 = 200\,\text{ft} \times 0.3627\,\text{ft}^3/\text{ft} = 72.54\,\text{ft}^3$$

   – Annular volume in the interval 1700 to 4450 ft plus an additional 25% as the casing is in the open wellbore

   $$V_2 = (4450 - 1600)\,\text{ft} \times 0.3132\,\text{ft}^3/\text{ft} \times 1.25 = 1115.775\,\text{ft}^3$$

   • Thus, the volume of cement required to secure the upper cementing interval is:

   $$V = V_1 + V_2 = 72.54 + 1115.775 = 1188.315\,\text{ft}^3$$

2) **The volume of cement required to secure the lower cementing interval (4450–4950 ft)**

   The volume of slurry required to cement the lower cementing interval will be the sum of the following volumes:
   • The annular volume in the interval 4450 to 4950 ft plus an additional 25% as the casing is in the open hole

   $$V_3 = (4950 - 4450)\,\text{ft} \times 0.3132\,\text{ft}^3/\text{ft} \times 1.25 = 195.8\,\text{ft}^3$$

   – Internal casing volume below the float collar

   $$V_4 = 42\,\text{ft} \times 0.4341\,\text{ft}^3/\text{ft} = 18.2\,\text{ft}^3$$

   Thus, the volume of cement required to secure the lower cementing interval is:

   $$V = V_3 + V_4 = 195.8 + 18.2 = 214.0\,\text{ft}^3$$

10.2.2 Volume of Displacing Fluid

$$V_{\text{d.f.}} = (4950 - 42)\,\text{ft} \times 0.0773\,\frac{\text{bbl}}{\text{ft}} = 379.4\,\text{bbl}$$

10.2.3 Pressure to Place the Cement Plug on the Stop Collar

It should be noted that as at this stage of the calculation, the hydrostatic pressures of the fluids are calculated, thus, the actual vertical depth data are used. As noted earlier, the pressure required for placing the cementing plug is equal to the difference between the hydrostatic pressures of the fluids in the annulus and inside the casing.

- **The hydrostatic pressure inside the casing.**
  - Drilling mud hydrostatic pressure:

  $$P_{dm} = 0.052 \times 11.5\,\text{ppg} \times (4950 - 42)\,\text{ft} = 2935\,\text{psi}$$

  - Hydrostatic pressure of cement slurry below the float collar:

  $$P_{c.s.} = 0.052 \times 16.4\,\text{ppg} \times 42\,\text{ft} = 35.8\,\text{psi}$$

  - The hydrostatic pressure inside the casing is equal to the sum of the hydrostatic pressures of the drilling mud and cement slurry below float collar:

  $$P_{B.K.} = 2935\,\text{psi} + 35.8\,\text{psi} = 2970.8\,\text{psi}$$

- **Hydrostatic pressure in the annulus.**

  The annulus in the example given contains drilling fluid up to 1400 m. However, as noted in the well information, 40 barrels of spacer were injected. This volume of spacer fluid is also in this interval. In order to calculate its hydrostatic pressure, it is first necessary to determine the interval it occupies in the annulus between two casings. The volume per unit length between the 13-3/8-in. and 9-5/8-in. casing is 0.0646 bbl/ft.

  Therefore, 40 barrels of buffer fluid in this annulus would occupy a section of length:

  $$40\,\text{bbl} \div 0.0646\,\frac{\text{bbl}}{\text{ft}} = 619.2\,\text{ft}$$

  The drilling mud accordingly occupies a section of the length:

  $$1400 - 619.2 = 780.8\,\text{ft}$$

  - The hydrostatic pressure of the drilling fluid in the annulus:

  $$P_{dm} = 0.052 \times 11.5\,\text{ppg} \times 780.8\,\text{ft} = 466.9\,\text{psi}$$

  - Hydrostatic pressure of the spacer solution in the annulus:

  $$P_{spacer} = 0.052 \times 12.5\,\text{ppg} \times 619.2\,\text{ft} = 402.5\,\text{psi}$$

- The hydrostatic pressure of the cement slurry in the annulus in the upper cementing interval:

$$P_{u.c.s.} = 0.052 \times 13.0\,\text{ppg} \times (4450 - 1400)\,\text{ft} = 2061.8\,\text{psi}$$

- The hydrostatic pressure of the cement slurry in the annulus in the lower cementing interval:

$$P_{l.c.s.} = 0.052 \times 16.4\,\text{ppg} \times (4950 - 4450)\,\text{ft} = 426.4\,\text{psi}$$

The hydrostatic pressure in the annulus will be equal to the sum of all the values calculated earlier:

$$P_{ann} = 466.9 + 402.5 + 2061.8 + 426.4 = 3357.6\,\text{psi}$$

The pressure required to place the cement plug on the stop collar (without taking into account friction pressure losses) $P_{s.c.}$ is

$$P_{s.c.} = 3357.6 - 2970.8 = 386.8\,\text{psi}$$

10.2.4 Buoyancy

Calculate the value of the buoyancy acting on the string using the borehole diagram shown in Figure 10.2.

Under static conditions, if the bottom casing arrangement includes a casing shoe with a check valve, the buoyancy force ΔF is calculated using the following equation:

$$\Delta F = \left[P_{ann} \times A_{OD}\right] - \left[m_{csg} + m_f\right]$$

where P_{ann} – hydrostatic pressure of fluids in the annulus (psi), A_{OD} cross-sectional area of the casing by outer diameter (square inch), m_{csg} – casing weight (lbm), m_f – mass of fluids in the casing (lbm).

If the lower casing assembly does not include a check-valve shoe, the ejection force ΔF is calculated using the following equation:

$$\Delta F = \left[P_{ann} \times (A_{OD} - A_{ID})\right] - \left\{(P_{ann} - P_{csg}) \times A_{ID}\right\} - m_{csg}$$

where P_{csg} – hydrostatic pressure of fluids in the casing (psi), A_{ID} casing cross-sectional area by inner diameter (square inch).

The pressure generated by the pump (P_{inj}) acting on the cross-sectional area at the inner diameter (A_{ID}) is also taken into account and the equation takes the form of

$$\Delta F = P_{BH} \times (A_{OD} - A_{ID}) + (P_{ID} \times A_{ID}) - m_{csg}$$

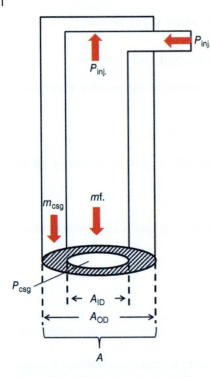

Figure 10.2 Schematic representation of the wellbore for calculating the value of the buoyancy force acting on the casing.

where P_{BH} – Bottom-hole pressure:

$$P_{BH} = P_{csg} + P_{inj}$$

If ΔF receives a positive value, the casing will rise out of the well. In practice, the maximum allowable injection pressure is the pressure at which $\Delta F = 0$. The task of the cementing team is not to exceed the maximum allowable injection pressure calculated according to the formula:

$$P_{max} = \frac{m_{csg}}{A_{OD}} - \left[P_{csg}\left(1 - \frac{A_{ID}}{A_{OD}}\right)\right]$$

where P_{max} – maximum allowable injection pressure.

Consider a calculation performed by the cementing crew to assess the risk of casing uplift using the following example: a 13-3/8-in. casing set at 700 ft depth, with a casing length unit weight of 61 lb/ft. and a cement slurry density of 14.8 lb/gall. Water with a density of 8.33 lb/gall was used as the displacing fluid.

1) Calculate the buoyancy force under static conditions:

$$\Delta F = \left\{(0.052 \times 14.8 \times 700)\left[\frac{\pi\left[(13.375)^2 - (12.515)^2\right]}{4}\right]\right\}$$
$$+ \left\{\left[(0.052 \times 14.8 \times 700) - (0.052 \times 8.33 \times 700)\right] \times \left[\frac{\pi(12.515)^2}{4}\right]\right\}$$
$$- (700 \times 61) = 9421 + 28971 - 42700 = -4308 \, lbf$$

A negative buoyancy force value indicates that the casing cannot be lifted under these conditions.

2) Calculate the pressure to place the cement plug:

$$P_{s.c.} = (14.8 - 8.33) \times 0.052 \times 700 = 235.5 \, psi$$

3) Calculate the maximum allowable injection pressure:

$$P_{max} = \left[\frac{700 \times 61}{(13.375)^2 \times \frac{\pi}{4}}\right] - \left[(700 \times 0.052 \times 8.33)\left(1 - \frac{12.515^2}{13.375^2}\right)\right]$$

$$= 303.91 - 37.74 = 266.17 \, lbf/in^2$$

This example uses the same fluid in the casing and in the annulus. In practice, there may be more than one such fluid and the calculations need to be modified accordingly.

10.3 Remedial Cementing Calculations

10.3.1 Plug Cementing Calculations

Cement plugs, especially plugs installed for sidetracking or to control influx, are usually hydrostatically balanced in the well, i.e. the hydrostatic pressure in the annulus and in the wellbore are equal at the time the plug is installed. The objective pursued by achieving such hydrostatic pressure equilibrium is to prevent a U-tube effect. In practice, this means that the casing serves as a kind of separator between the same fluid in the casing and the annulus. In order to maintain the same hydrostatic pressure, these fluids naturally have to occupy the same length of casing both inside and in the annulus.

Let us look at the basic formulas used in calculations during the cement plug installation process.

1) Volume of cement slurry, V_{cement}:

$$V_{cement} = L \times S_{o.h.}$$

where L – length of the cement column in the open hole (ft), $S_{o.h.}$ – volume per unit length of open hole in ft^3/ft (this parameter is listed in various industry guides).

2) Cement plug length with the casing in place, L_{PL}:

$$L_{PL} = \frac{V_{cement}}{S_{ann} + S_{tub}}$$

where S_{ann} – volume per unit length of annular space between the string and the open hole (ft^3/ft), S_{tub} – volume per unit length of string (ft^3/ft).

3) Volume of spacer injected after the cement slurry, V_{spc2}:

$$V_{spc2} = \frac{V_{spc1}}{S_{ann}} \times S_{tub}$$

where V_{spc1} – volume of spacer pumped before cement slurry.

4) Volume of displacing fluid:

$$V_{disp.fl.} = S_{tub} \times \left[D - \left(L_{LP} + L_{spc2} \right) \right]$$

where D – length of the string (to the end of the cement plug bridge) (ft), $L_{буф.2.}$ – height of spacer column injected after the cement slurry, calculated as:

$$L_{spc2} = \frac{V_{spc2}}{S_{tub}}$$

Consider the calculation carried out by the cementing crew when installing a cement plug in the following example (Figure 10.3):

Well data
- Open hole diameter: 8-1/2 in.
- Volume per unit length of open hole ($S_{o.h.}$) – 0.3941 ft³/ft
- Drill string:
  - Diameter – 4 in.
  - Weight per unit length of the string (S_{tub}) 14.0 lbm/ft
  - Volume per unit length of drill string $S_{d.s.}$ – 0.01084 bbl/ft or 0.06084 ft³/ft
- Volume per unit length of annular space (S_{ann}) – 0.0546 bbl/ft or 0.3068 ft³/ft
- Volume of spacer pumped before cement slurry V_{spc1} – 10 bbl

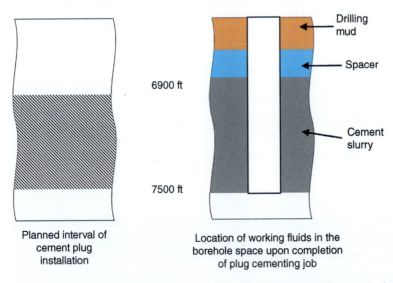

Figure 10.3 Schematic representation of the wellbore for calculations made during the installation of the cement plug.

1) Calculate the volume of cement slurry:

$$V_{cem} = L \times S_{o.h.} = 600\,\text{ft} \times 0.3941\,\text{ft}^3/\text{ft} = 236.46\,\text{ft}^3$$

2) Find the length of the cement plug with casing in place, L_{PL}:

$$L_{PL} = \frac{V_{cem}}{S_{ann} + S_{tub}} = \frac{236.46\,\text{ft}^3}{0.3068\,\text{ft}^3/\text{ft} + 0.06084\,\text{ft}^3/\text{ft}} = 643.2\,\text{ft}$$

3) Volume of spacer injected after the cement slurry, V_{spc2}:

$$V_{spc2} = \frac{V_{spc1}}{S_{ann}} \times S_{tub} = \frac{10\,\text{bbl}}{0.0546\,\text{bbl}/\text{ft}} \times 0.01084\,\text{bbl/ft}$$

4) Volume of displacing fluid:

$$V_{displ.fl.} = S_{tub} \times \left[D - \left(L_{PL} + L_{spc2}\right)\right] = 0.01084\,\frac{\text{bbl}}{\text{ft}}$$
$$\times \left[7500\,\text{ft} - \left(643.2\,\text{ft} + \frac{2.0\,\text{bbl}}{0.01084\,\text{bbl}/\text{ft}}\right)\right] = 72.3\,\text{bbl}$$

Another example of calculations for cement-plug installation:

For cement plug installation calculations, it is convenient to use schematic diagrams of a well, and two diagrams should be made with and without a casing in place (Figure 10.4). The volume of cement slurry is calculated using a diagram without casing. Consider a well with the following data:

- True vertical depth 6000 ft
- Casing
  - Diameter 8-5/8 in.
  - Weight per unit length 38 lbm/ft
- Work string
  - Diameter 4 in.
  - Weight per unit length 14 lbm/ft
- Cement slurry
  - Mix water 4.3 gal/sk
  - Slurry yield 1.06 ft³/sk
- Height of cement plug – 300 ft

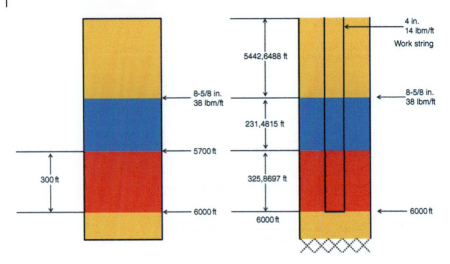

Figure 10.4 Schematic representation of the wellbore for calculations made during the installation of the cement plug.

1) Volume of cement slurry:

    $300 \text{ ft} \times 0.0587 \text{ bbl/ft} = 17.61 \text{ bbl}$

2) Number of sacks of cement required to prepare a cement slurry of 17.61 bbl:

    $17.61 \text{ bbl} \div 0.1781 \text{ bbl/ft}^3 \div 1.06 \text{ ft}^3/\text{sk} = 93.2802 \text{ sk}$

3) Volume of mixing water:

    $93.2802 \text{ sk} \times 4.3 \text{ gal/sk} \div 42 \text{ gal/bbl} = 9.5501 \text{ bbl}$

4) Cement plug lifting height:

    $$\begin{array}{r} 6000 \text{ ft} \\ -3000 \text{ ft} \\ \hline 5700 \text{ ft} \end{array}$$

5) With the working casing running downhole, a new capacity factor (K_C) is introduced. This factor is equal to the sum of the volume per unit length of annular space between two casings (V_{ann}) and the volume per unit length of empty casing (V_{cas}). All these data are available in technical manuals for drilling engineers.

    $V_{ann} \text{ bbl/ft} + V_{cas} \text{ bbl/ft} = K_C \text{ bbl/ft}$
    $0.0432 \text{ bbl/ft} + 0.01084 \text{ bbl/ft} = 0.05404 \text{ bbl/ft}$

6) Cement plug lift height including capacity factor

   17.61 bbl ÷ 0.05404 bbl/ft = 325.8697 ft

7) Lifting height of 10 barrels of a spacer in the annulus

   10.0 bbl ÷ 0.0432 bbl/ft = 231.4815 ft

8) Required volume of spacer to fill 231.4815 ft of casing

   231.4815 ft × 0.01084 bbl/ft = 2.5093 bbl

9) Displacement fluid lifting height

   $$\begin{array}{r} 6000.0000\,\text{ft} \\ -325.8697\,\text{ft} \\ \underline{-231.4815\,\text{ft}} \\ 5442.6488\,\text{ft} \end{array}$$

10) Displacement fluid volume for 5442.6488 ft of casing

    5442.6488 ft × 0.01084 bbl/ft = 58.9983 bbl

Thus, the fluid injection scheme for a cement plug bridge installation would look as follows:

- The well is circulated
- 10 barrels of buffer fluid/water are pumped
- 17.61 barrels cement slurry will be pumped
- 2.5093 barrels of buffer liquid/water after the cement slurry is pumped
- 58.9983 barrels of displacement fluid are pumped

10.3.2 Squeeze Cementing

Two main types of calculations are carried out in squeeze cementing:

- Calculations of different volumes:
  - Work casing volume
  - Casing volume below the working casing string
  - Displacing fluid volume
- Pressure calculations at various points in the wellbore at various stages of the process:
  - Pressure to kill the well
  - Injection pressure
  - Bottomhole pressure (at different stages of the process)
  - Maximum allowable wellhead pressure

230 | 10 Typical Calculations for Well Cementing

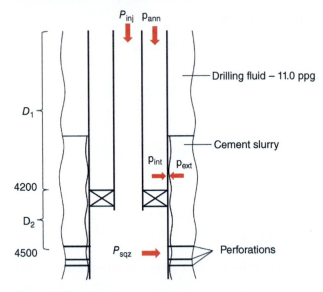

Figure 10.5 Schematic representation of the wellbore for calculations during the squeeze cementing process.

This list does not include all parameters that are calculated when carrying out squeeze cementing.

It should be noted that the first group uses the measured depth for calculations and the true vertical depth is used for pressure calculations. Ideally, during this type of operation, it is desirable to avoid fracturing and the maximum pressure achieved during the operation exceeds the injection pressure by 500 or 1000 psi.

Let us look at the calculations performed while squeeze cementing, using the well shown in Figure 10.5 as an example.

The most vulnerable part of the casing is the interval opposite the packer. Increased (P_{ext}) annulus pressure in this section between the casing and the formation can lead to the collapse of the casing. The P_{ext} is made up of the hydrostatic pressure of the fluids in the annulus and the injection pressure. In case the annulus is separated from the wellhead (e.g. by means of a cement plug) or is blocked at the wellhead, the P_{ext} is calculated according to the formula:

$$P_{ext} = P_{inj} + \left[0.052 \times (D_1 + D_2) \times \rho_1\right] - \left(0.052 \times D_2 \times \rho_2\right)$$

10.3 Remedial Cementing Calculations

where D_1 – depth of the packer, D_2 – distance from perforations to the packer, P_{inj} – maximum injection pressure (maximum pump pressure), ρ_1 – density of the fluid inside the string, and ρ_2 – density of the lightest fluid pumped during cementing.

When there is more than one fluid in the annulus and casing, the sum of all hydrostatic pressures of the individual fluids must be taken into account.

If annulus is not separated from wellhead, P_{ext} is calculated by formula:

$$P_{ext} = \left(0.052 \times D_1 \times \rho_1\right)$$

- D_1 – depth of the packer
- D_2 – distance from perforations to the packer
- P_{ann} – wellhead pressure applied to the annular space between the workstring and casing
- P_{ext} – pressure in the annular space between the casing and the formation at the packer level
- P_{int} – pressure in the annular space between the workstring and casing at the packer level
- P_{inj} – injection pressure/pump pressure
- P_{sqz} – the squeeze pressure, i.e. the pressure that the reservoir experiences during the squeeze cementing process.

Consider a situation where cementing operations in the aforementioned well reach an injection pressure of 4000 psi and the completion fluid density is 8.5 ppg. Given that the packer is at a depth of 4200 m, let us calculate P_{ext}:

$$P_{ext} = P_{inj} + \left[0.052 \times (D_1 + D_2) \times \rho_1\right] - \left(0.052 \times D_2 \times \rho_2\right)$$
$$= 4000 + \left(0.052 \times 4500 \times 8.5\right) - \left(0.052 \times 300 \times 8.5\right) = 5856 \text{ psi}$$

For a casing collapse pressure of 4910 psi, the pressure inside the casing against the packer must be at least 946 psi. The hydrostatic pressure generated by an 8.5 ppg column of completion fluid at the packer level is

$$P_h = 0.052 \times 4200 \times 8.5 = 1856 \text{ psi}$$

Thus, the hydrostatic pressure opposite the packer exceeds the minimum allowable pressure by 910 psi to prevent the collapse of the string and ensure that the cementing operation is conducted safely.

Let us consider another example of a typical calculation for squeeze cementing (Figure 10.6).

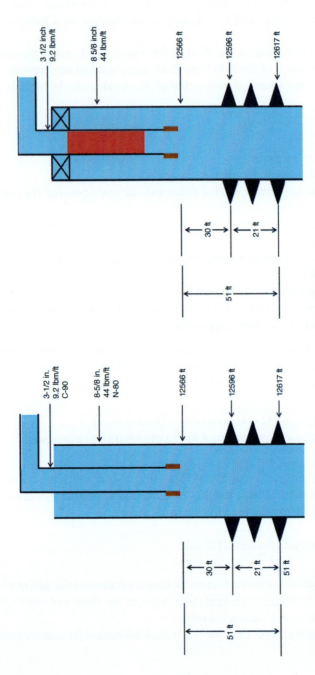

Figure 10.6 Schematic representation of the wellbore for calculations during the squeeze cementing process.

10.3 Remedial Cementing Calculations

Casing
- Outer diameter 8-5/8 in.
- Weight per length 44 lbm/ft
- Steel grade N-80
- Collapse pressure 6950 psi
- Yield point 8120 psi
- Volume per unit of length 0.0564 bbl/ft

Cement slurry
- Number of sacks 115
- Density 15.4 ppg
- Slurry yield 1.24 ft^3/sk
- Mix water 5.6 gal/sk
- Volume of slurry 25.3971 bbl
- Hydrostatic pressure per unit length 0.8 psi/ft

Workstring
- Outer diameter 3-1/2 in.
- Inner diameter 2.992 in.
- Weight per unit length 9.2 lbm/ft
- Steel grade C-90
- Collapse pressure 10380 psi
- Yield point 11430 psi
- Volume per unit length 0.0087 bbl/ft
- Volume per unit length of annular space between workstring and casing 0.0446 bbl/ft

Additional information
- Packer depth 12 566 ft
- Perforation interval 12 596 12 617 ft
- Fracture gradient 1.02 psi/ft
- Completion fluid density 8.6 ppg
- Annulus volume per unit length 0.0564 bbl/ft
- Volume of displacing fluid 20 bbl

1) Calculation of volumes:
   - Mix water (115 sk)
     115 sk × 5.6 gal/sk ÷ 42 gal/bbl = 15.3333 bbl

   - Volume of cement slurry (115 sk)
     115 sk × 1.24 ft^3/sk ÷ 0.1781 bbl/ft^3 = 25.3971 bbl

   - Volume of workstring (12566 ft)
     12566 ft × 0.0087 bbl/ft = 109.3242 bbl

   - Casing volume up to perforation interval (30 ft)
     30 ft × 0.0564 bbl/ft = 1.692 bbl

   - Casing volume within the perforation interval (21 ft)
     21 ft × 0.0564 bbl/ft = 1.1844 bbl

- Casing volume to the end of the perforation interval:

  1.692 bbl + 1.1844 bbl = 2.8764 bbl

- Volume of the annulus up to the packer:

  12566 ft × 0.0446 bbl/ft = 560.4436 bbl

2) Calculate the pressure required to circulate one barrel of cement slurry back through the workstring:

   The hydrostatic pressure of the fluid per unit length is determined by the formula:

   $P_{hyd} = 0.052 \times \rho_{fluid} \times 1\,ft$

   Cement $P_{hyd} = 0.052 \times \rho_{fluid} \times 1\,ft = 0.052 \times 15.4 \times 1 = 0.8\,psi/ft$

- Completion fluid $P_{hyd} = 0.052 \times \rho_{fluid} \times 1\,ft = 0.052 \times 8.6 \times 1 = 0.44\,psi/ft$

  0.80 psi/ft (cement)

  −0.44 psi/ft (completion fluid)

  0.35 psi/ft (differential pressure)

Pressure required to circulate one barrel of cement slurry back through the workstring:

0.35 psi/ft ÷ 0.0087 bbl/ft = 40.59 psi/bbl

The pressure required for reverse circulation of the entire volume of cement slurry is:

25.3971 bbl × 40.59 psi/bbl = 1031 psi

3) Calculate the maximum wellhead pressure (squeeze pressure):

   The height of 25.3971 barrels of cement slurry is:

   25.3971 bbl ÷ 0.0087 bbl/ft = 2.919 ft

   Height of completion fluid:

   12 596.0000 ft

   −2919.2069 ft

   9676.7931 ft

   Hydrostatic pressure up to the perforation interval:

   9676.7931 ft × 0.4468 psi/ft =   4323.5912 psi

   +2919.2069 ft × 0.8000 psi/ft = +2335.3655 psi

   12596.0000 ft   6658.9567 psi

Fracture pressure (minus a safety factor of 250 psi)

12596.0000 ft × 1.02 psi/ft = 12847.92 psi
12847.92 psi − 250 psi = 12597.92 psi

Calculate the maximum wellhead pressure

12597.9200 psi
−6658.9567 psi
5938.9633 pssi

4) Calculate the casing collapse pressure at the determined squeeze pressure of 5938.9633 psi (minus a safety factor of 250 psi).
   - For calculations, we take a casing collapse safety factor of 65%

   6950 psi × 65% = 4517.5 psi

   4517.5 psi − 250 psi = 4267.5 psi

   - The hydrostatic pressure in the annulus up to the packer is:

   12566 ft × 0.4468 psi/ft = 5614.4888 psi

   Note also that in the annulus, the "trapped" pressure is 1031.0638 psi
   - The sum of all pressures listed here shows the casing collapse pressure

   4267.5 psi
   +5614.4888 psi
   +1031.0638 psi
   10913.0526 psi

5) Calculate the casing collapse pressure at the determined squeeze pressure of 5938.9633 psi (minus a safety factor of 250 psi).
   - Hydrostatic pressure in the workstring acting on the packer:

   2919.2069 ft × 0.8 psi/ft = 2335.3655 psi
   +9646.7931 ft × 0.4468 psi/ft = +4310.1872 psi
   12566 ft   6645.5527 psi

   5938.9633 psi − pressure at the wellhead (squeeze)
   +6645.5527 psi − hydrostatic pressure in the workstring
   12584.5160 psi − pressure aimed for casing collapse

Thus, the pressure to collapse the column exceeds the column's collapse resistance by 1671.4634 psi:

$$12584.5160\,psi - 10913.0526\,psi = 1671.4643\,psi$$

In this case, there are two scenarios:

1) Reduce the maximum squeeze pressure by 1671.4643 psi

> 5938.9633
> −1671.4634
> 4267.4999 psi – maximum squeeze pressure

2) Increase the collapse resistance value for the casing by increasing the annular pressure by 1671.4643 psi

> 1031.0638 psi – current wellhead pressure in the annulus
> +1671.4643 psi – additional pressure in the annulus
> 2702.5272 psi – pressure in the annulus required to prevent casing collapse

Annex. Conversion Tables

Oil and Gas Well Cementing for Engineers, First Edition. Baghir A. Suleimanov, Elchin F. Veliyev, and Azizagha A. Aliyev.
© 2023 John Wiley & Sons Ltd. Published 2023 by John Wiley & Sons Ltd.

Length

Convert to:

Convert from:	micrometer	millimeter	centimeter	meter	kilometer	inch	foot	yard	mile (SI)	Nautical mile (SI)	Nautical mile (Imperial)
micrometer	1	10^{-3}	10^{-4}	10^{-6}	10^{-9}	3.937×10^{-5}	3.281×10^{-6}	1.094×10^{-6}	6.214×10^{-10}	5.399×10^{-10}	5.396×10^{-10}
millimeter	1000	1	0.1	0.001	10^{-6}	3.937×10^{-2}	3.281×10^{-3}	1.094×10^{-3}	6.214×10^{-7}	5.399×10^{-7}	5.396×10^{-7}
centimeter	10^{4}	10	1	0.01	10^{-5}	0.3937	3.281×10^{-2}	1.094×10^{-2}	6.214×10^{-6}	5.399×10^{-6}	5.396×10^{-6}
meter	10^{6}	1000	100	1	1000	39.37	3.2808	1.0936	6.214×10^{-4}	5.399×10^{-4}	5.396×10^{-4}
kilometer	10^{9}	10^{6}	10^{5}	1000	1	3.937×10^{4}	3.2808×10^{3}	1.0936×10^{3}	0.621371	0.539957	0.539612
inch	25400	25.4	2.54	0.0254	2.54×10^{-5}	1	0.0833	0.0278	1.578×10^{-5}	1.372×10^{-5}	1.371×10^{-5}
foot	304800	304.8	30.48	0.3048	3.048×10^{-4}	12	1	0.3333	1.894×10^{-4}	1.646×10^{-4}	1.645×10^{-4}
yard	914400	914.4	91.44	0.9144	9.144×10^{-4}	36	3	1	5.682×10^{-4}	4.937×10^{-4}	4.934×10^{-4}
mile (SI)	1.609×10^{9}	1.609×10^{6}	1.609×10^{5}	1609.344	1.609344	63360	5280	1760	1	0.86898	0.86842
Nautical mile (SI)	1.852×10^{9}	1.852×10^{6}	1.852×10^{5}	1852	1.852	72913.386	6076.115	2025.372	1.1507794	1	0.9993611
Nautical mile (Imperial)	1.853×10^{9}	1.853×10^{6}	1.853×10^{5}	1853.184	1.853184	72960	6080	2026.667	1.1515152	1.000639	1

Weight

Convert from:	milligram	gram	kilogram	ton	pound	ounce	pood
milligram	1	0.001	10^{-6}	10^{-9}	2.205×10^{-6}	3.5274×10^{-5}	6.1047×10^{-8}
gram	1000	1	0.001	10^{-6}	2.2046×10^{-3}	3.527396×10^{-2}	6.1047×10^{-5}
kilogram	10^6	1000	1	0.001	2.20462	35.27396195	6.1047×10^{-2}
ton	10^9	10^6	1000	1	2.204622622×10^3	3.527396195×10^4	61.0475
pound	4.53592×10^5	4.53592×10^2	0.453592	4.53592×10^{-4}	1	16	0.0276907
ounce	2.83495×10^4	28.3495	2.83495×10^{-2}	2.83495×10^{-5}	0.0625	1	1.73067×10^{-3}
pood	1.63807×10^7	1.63807×10^4	16.3807	1.63807×10^{-2}	36.1132	577.812	1

Volume

Convert from:	Convert to: cubic millimeter	cubic centimeter	cubic meter	milliliter	liter	oil barrel	barrel (US)	barrel (UK)	gallon (US)	gallon (imperial)	cubic feet	cubic inches
cubic millimeter	1	0.001	10^{-9}	0.001	10^{-6}	6.29×10^{-9}	8.386×10^{-9}	6.11×10^{-9}	2.642×10^{-7}	2.199×10^{-7}	3.531×10^{-8}	6.102×10^{-5}
cubic centimeter	1000	1	10^{-6}	1	0.001	6.29×10^{-6}	8.386×10^{-6}	6.11×10^{-6}	2.642×10^{-4}	2.199×10^{-4}	3.531×10^{-5}	6.102×10^{-2}
cubic meter	10^9	10^6	1	10^6	1000	6.2898	8.3864	6.1103	264.172	219.969	35.315	6.102×10^{-4}
milliliter	1000	1	10^{-6}	1	0.001	6.29×10^{-6}	8.386×10^{-6}	6.11×10^{-6}	2.642×10^{-4}	2.199×10^{-4}	3.531×10^{-5}	6.102×10^{-2}
liter	10^6	1000	0.001	1000	1	6.29×10^{-3}	8.386×10^{-3}	6.11×10^{-3}	0.2642	0.21997	3.531×10^{-2}	61.024
oil barrel	1.59×10^8	1.59×10^5	0.15899	1.59×10^5	1.59×10^2	1	1.333	0.9715	42	34.972	5.615	9702
barrel (US)	1.192×10^8	1.192×10^5	0.1192	1.192×10^5	1.192×10^2	0.75	1	0.7286	31.5	26.229	4.211	7276.5
barrel (UK)	1.637×10^8	1.637×10^5	0.1637	1.637×10^5	1.637×10^2	1.029	1.3725	1	43.234	36	5.78	9987.1
gallon (US)	3.785×10^6	3.785×10^3	3.785×10^{-3}	3.785×10^3	3.785	0.0238	0.0317	0.0231	1	0.8327	0.1337	231
gallon (imperial)	4.546×10^6	4.546×10^3	4.546×10^{-3}	4.546×10^3	4.546	2.859×10^{-2}	3.813×10^{-2}	2.778×10^{-2}	1.201	1	0.1605	277.419
cubic feet	2.832×10^7	2.832×10^4	2.832×10^{-2}	2.832×10^4	28.317	0.1781	0.2375	0.17302	7.4805	6.2288	1	1728
cubic inches	1.639×10^4	16.387	1.639×10^{-5}	16.387	1.639×10^{-2}	1.031×10^{-4}	1.374×10^{-4}	1.001×10^{-4}	4.329×10^{-3}	3.605×10^{-3}	5.787×10^{-4}	1

Pressure

Convert to:

Convert from:	pascal	kilopascal	megapascal	N/m²	bar	pound per square inch	millimeter of mercury	atm	millimeter of water
pascal	1	0.001	10^{-6}	1	10^{-5}	1.45×10^{-4}	7.501×10^{-3}	9.869×10^{-6}	0.102
kilopascal	1000	1	0.001	1000	0.01	0.14504	7.5006	9.869×10^{-3}	101.974
megapascal	10^6	1000	1	10^6	10	145.038	7500.638	9.8692	1.0197×10^5
N/m²	1	0.001	10^{-6}	1	10^{-5}	1.45×10^{-4}	7.501×10^{-3}	9.869×10^{-6}	0.102
bar	10^5	100	0.1	10^5	1	14.5038	750.064	0.9869	1.0197×10^4
pound per square inch	6894.757	6.895	6.895×10^{-3}	6894.757	0.06895	1	51.7151	0.06805	703.089
millimeter of mercury	133.322	0.133322	1.333×10^{-4}	133.322	1.333×10^{-3}	0.01934	1	1.316×10^{-3}	13.595
atm	1.013×10^5	101.325	0.1013	1.013×10^5	1.01325	14.696	760	1	1.033×10^4
millimeter of water	9.8064	9.8064×10^{-3}	9.8064×10^{-6}	9.8064	9.8064×10^{-5}	1.422×10^{-3}	0.07355	9.678×10^{-5}	1

Area

Convert from:	Convert to:							
	square millimeter	square centimeter	square meters	square kilometers	hectare	square inches	square feet	square yard	acre
square millimeter	1	0.01	10^{-6}	10^{-12}	10^{-10}	1.55×10^{-3}	1.076×10^{-5}	1.196×10^{-6}	2.471×10^{-10}
square centimeter	100	1	10^{-4}	10^{-10}	10^{-8}	0.155	1.076×10^{-3}	1.196×10^{-4}	2.471×10^{-8}
square meters	10^6	10^4	1	10^{-6}	10^{-4}	1550	10.7639	1.19599	2.471×10^{-4}
square kilometers	10^{12}	10^{10}	10^6	1	100	1.55×10^9	1.076×10^7	1.196×10^6	247.105
hectare	10^{10}	10^8	10^4	0.01	1	1.55×10^7	1.076×10^5	1.196×10^4	2.47105
square inches	645.16	6.4516	6.4516×10^{-4}	6.4516×10^{-10}	6.4516×10^{-8}	1	6.944×10^{-3}	7.716×10^{-4}	1.594×10^{-7}
square feet	9.290×10^4	929.03	9.29×10^{-2}	9.29×10^{-8}	9.29×10^{-6}	144	1	0.1111	2.296×10^{-5}
square yard	8.361×10^5	836.127	0.8361	8.361×10^{-7}	8.361×10^{-5}	1296	9	1	2.066×10^{-4}
acre	4.047×10^9	4.047×10^7	4046.86	4.047×10^{-3}	0.4047	6.273×10^6	43560	4840	1

Density

Convert from:	Convert to:					
	g/cm³	kg/m³	g/l	lb/in³	lb/ft³	lb/gal (imperial)	lb/gal (US)
g/cm³	1	1000	1000	0.0361	62.428	8.3454	10.022
kg/m³	0.001	1	1	3.613×10^{-5}	6.243×10^{-2}	0.01	0.0083
g/l	0.001	1	1	3.613×10^{-5}	6.243×10^{-2}	0.01	0.0083
lb/in³	27.7	27680	27680	1	1728	277	231
lb/ft³	0.016	16	16	5.79×10^{-4}	1	0.161	0.134
lb/gal (imperial)	0.0998	99.8	99.8	3.6×10^{-3}	6.23	1	0.833
lb/gal (US)	0.12	120	120	4.33×10^{-3}	7.48	1.2	1

Recommended Literature

1 Bulatov A. I., Savenok O. V. Zakanchivanie neftianykh i gazovykh skvazhin: teoriya i praktika. [Completion of Oil and Gas Wells: Theory and Practice]. Prosveshenie-Yg. 2010.
2 Oatman F. W. (1915). Water intrusion and methods of prevention in california oil fields. Trans 48: 627–650. doi: 10.2118/915627-G.
3 Edwards G. C., Angstadt R. L. (1966). The effect of some soluble inorganic admixtures on the early hydration of portland cement. Journal of Applied Chemistry 16 (5): 166–168.
4 Suleymanov A. B., Karapetov K. A., Yashin A. S. Prakticheskie raschety pri tekushchem i kapital'nom remonte skvazhin [Practical Calculations for Well Servicing and Workover]. Nedra. 1984.
5 Mirzadzhanzade A. K., Entov V. M. Gidrodinamika v bureni [Hydrodynamics in Drilling]. Nedra. 1985.
6 Mil'shteyn V. M. Tsementirovanie burovykh skvazhin [Well cementing], Krasnodar. 2003.
7 Sazonov A. A. Tsementirovanie neftianykh i gazovykh skvazhin [Cementing of Oil and Gas Wells]. ChentrLitNefteGas. 2010.
8 Nelson E. B. (ed.). Well cementing. Newnes. 1990.
9 Liu G. (ed.). Applied well cementing engineering. Gulf Professional Publishing. 2021.
10 Crook R. Cementing horizontal wells. Halliburton. 2008.
11 Mitchell R. F. Petroleum engineering handbook, Drilling Engineering, vol. II. Society of Petroleum Engineers. 2007.
12 Guo B., Liu G. Applied drilling circulation systems: hydraulics, calculations and models. Gulf Professional Publishing. 2011.
13 Barnes P., Bensted J. Structure and performance of cements. CRC Press. 2002.

Oil and Gas Well Cementing for Engineers, First Edition. Baghir A. Suleimanov, Elchin F. Veliyev, and Azizagha A. Aliyev.
© 2023 John Wiley & Sons Ltd. Published 2023 by John Wiley & Sons Ltd.

14 Caenn R., Darley H. C. H., Gray G. R. Composition and properties of drilling and completion fluids. Gulf professional publishing. 2011.
15 Watters L. T., Dunn-Norman S. Petroleum well construction. Wiley. 1998.
16 Kurdowski W. Cement and concrete chemistry. Springer Science & Business. 2014.

Index

a

absolute volume 208, 209
accelerators
    calcium chloride ($CaCl_2$) 32–34
    calcium formate ($Ca(HCOO)_2$) 35
    calcium nitrate ($Ca(NO_3)_2$) 36
    calcium nitrite ($Ca(NO_2)_2$) 36
    chlorine anions 35
    definition 32
    inorganic salts 32
    sodium chloride 32, 34
    triethanolamine ($N(C_2H_4OH)_3$) 36
acids and salts 39
acid-soluble cements 72
acoustic logging 173–176
additive concentration calculation 209–210
adsorption theory 36
American Petroleum Institute (API) classification
    casing steel grades
        C-90 steel grade 90, 91
        H-40 grade steel 89, 90
        J-55 grade 89, 90
        K-55 grade 89, 90
        L-80 grade steel 90, 91
        N-80 grade steel 90, 91
        T-95 steel grade 90, 91
    cement slurry preparation calculations 207, 208

of Portland cement
    class A cement 25, 26
    class B cement 25, 26
    class C cement 25, 27
    class D cement 25, 27
    class E cement 25, 27
    class F cement 25, 28
    class G cement 25, 28
    class H cement 25, 28–29
amorphous condensed microsilica 45–46
amplitude map (cement map) 177
antifoaming agents (defoamers) 55–56
API classification *see* American Petroleum Institute (API) classification
API plastic collapse pressure P_P 94
API RP 10B 16
API spec 10A 16
API transition collapse pressure P_{tr} 92, 93
API yield collapse pressure 94
appraisal well 5
artificial pozzolans 43, 44
atmospheric consistometers 182, 183
Azerbaijan Scientific Research Petroleum Institute (AzNII) 197, 198
"AzNII cone" 197, 198

b

ball-type check valve 99, 100
barite 50

Oil and Gas Well Cementing for Engineers, First Edition. Baghir A. Suleimanov, Elchin F. Veliyev, and Azizagha A. Aliyev.
© 2023 John Wiley & Sons Ltd. Published 2023 by John Wiley & Sons Ltd.

Index

basket cementing 12–13
bentonite 40–42, 216–218
BFS systems *see* Blast Furnace Slag (BFS) systems
BHA *see* bottom hole assembly (BHA)
BHCT *see* bottom hole circulating temperature (BHCT)
BHST *see* bottom hole static temperature (BHST)
Bingham fluids 60
Bingham model 192–194
Blast Furnace Slag (BFS) systems 66–67
blind valves 99
borehole
  axis 1, 2
  buoyancy 223, 224
  curvature of 3, 4
  definition 1
  radioactive tracers 56
  temperature log 168
  turbulators 102
bottom hole assembly (BHA) 1
bottom hole circulating temperature (BHCT) 113–114, 139
bottom hole static temperature (BHST) 114, 139
bow-string centralizers 101
bulk volume 208–209
burst pressure 91
butterfly valves 99

C

calcium aluminate cements 63
calcium chloride ($CaCl_2$) 32–34
calcium formate ($Ca(HCOO)_2$) 35
calcium hydrosilicate (C–S–H) 20
calcium nitrate ($Ca(NO_3)_2$) 36
calcium nitrite ($Ca(NO_2)_2$) 36
caliper 110–112
carboxymethylhydroxyethylcellulose (CMHEC) 38
cased-hole remedial cementing equipment 106–108
casing
  connection types of 95–96
  hardware
    casing shoe 96–99
    cementing basket 105, 106
    cementing head 104, 105
    cementing plugs 103–104
    centralizers 100–102
    check valve 99–100
    scratcher 103
    screening devices 105, 106
    turbulator 102, 103
  steel grades
    C-90 steel grade 90, 91
    H-40 grade steel 89, 90
    J-55 grade 89, 90
    K-55 grade 89, 90
    L-80 grade steel 90, 91
    N-80 grade steel 90, 91
    T-95 steel grade 90, 91
  strength characteristics of
    burst pressure 92
    collapse pressure 92–94
    yield strength 91
  types of
    conductor casing 86
    diameter 85
    intermediate 86
    liner 87–88
    production 86
    surface 86
  weight per unit length of tube 94–95
casing collar locator (CCL) 176
casing shoe
  definition 96–97
  guide shoe 97, 98
  lipstick shoe 98
  reamer shoe 97, 98
  self-orienting shoe 98
Casson model 195
cavernometry 110
CCL *see* casing collar locator (CCL)
cellulose derivatives 38
cement additives
  accelerators
    calcium chloride ($CaCl_2$) 32–34
    calcium formate ($Ca(HCOO)2$) 35
    calcium nitrate ($Ca(NO_3)_2$) 36
    calcium nitrite ($Ca(NO_2)_2$) 36

chlorine anions   35
definition   32
inorganic salts   32
sodium chloride   32, 34
triethanolamine (N(C$_2$H$_4$OH)$_3$)   36
antifoaming agents (defoamers)
  55–56
cement slurry viscosity modifiers
  (dispersants)
  definition   51
  plasticizers   51
  rheological properties of   50–51
  sedimentation   52–53
  superplasticizers   51
  water separation   52
extenders   31
  cement slurry density reduction of   39
  cement slurry volume, increase of   39
  clay minerals   40, 42
  extenders with low density   40
  gas-based   40, 48
  pozzolans   43–46
  sodium silicate   43
fluid loss agents   32
  particulate materials   54
  water soluble polymers   54–55
lost circulation prevention agents
  32, 55
mud decontamination   57
radioactive tracers   56–57
retarders   31
  adsorption theory   36
  cellulose derivatives   38
  hydroxycarboxylic acid   38
  inorganic compounds   39
  lignosulfonates   37–38
  organophosphonates   39
  precipitation theory   36
  saccharide compounds   38
  theory of complexation   36
  theory of nucleation   36
strengthening agents   56
weighting agents   31, 48–50
cement bond log (CBL)   176
cementing basket   105, 106
cementing equipment
  casing (see casing)
  remedial cementing equipment
    cased-hole   106–108
    open hole   108
  surface equipment
    cement trucks   78
    displacement tank system   79
    hydraulic mixer   81–82
    liquid additive dosing systems   80
    liquid additive metering system   80
    on-the-fly mixing   79
    pneumatic loading (dry material)   75, 77
    pneumatic mixing tank   75, 77
    premixing liquid additives with
      water   79–80
    recirculation hydraulic mixer   82–84
    screw-type unloader   75, 77
    storage silo   75, 76
    storage, transportation, and preparation
      of   75, 76
    surge tank   81
cementing head   104, 105, 135, 136
cementing plugs   14, 15, 103–104
cement job evaluation
  acoustic logging   172–176
  hydraulic testing
    inflow test   167
    pressure test   164–167
  logging tools, types of
    cement bond log   176
    multiple pad sonic tool   177
    radial acoustic cement meter   177
    ultrasonic tool   177–178
  radioactive logging   169–171
  temperature log   167–169
cement plug installation
  process   225
  with the use of coiled tubing   146–147
  using a dump bailer   145–146
  using the two plugs method   146
cement slurry
  chemical analysis of mix water   205
  preparation calculations
    absolute volume   208, 209
    additive concentration
      calculation   209–210

cement slurry (cont'd)
   API classification   207, 208
   bentonite   216–218
   bulk volume   208–209
   density and yield of   210–212
   fly ash   214–216
   sodium salts   212–214
   specific gravity of cement slurry   208
   weighting agents   218
  preparation of   180–181
  test methods of
   density   181–182
   flowability of cement slurries 197–199
   fluid loss   186, 187
   free water   187–188
   rheological measurements   188–196
   sedimentation test   188
   static gel strength   196–197
   thickening time   182–186
cement slurry circulation disruption   55
cement slurry mixers (FANN Instruments)   180
cement slurry viscosity modifiers (dispersants)
  definition   51
  plasticizers   51
  rheological properties of   50–51
  sedimentation   52–53
  superplasticizers   51
  water separation   52
cement stone test methods
  destructive test (compressive strength)   199–201
  expansion and shrinkage   200
  gas migration   202
  mechanical strength of cement   199
  permeability   202
  thermophysical properties of cement
   coefficient of linear thermal expansion   203, 204
   thermal conductivity   203
cement trucks   78
centralizers   100–102
ceramic microspheres   47–48

check valve   99–100
chemical composition of Portland cement
  dicalcium silicate (C_2S/belite)   19, 20
  hydration process of   21–22
  tetracalcium alumoferrite (C_4AF/braunmillerite)   20
  tricalcium aluminate (C_3A)   20
  tricalcium silicate (C_3S/alite)   19, 20
chemically bonded phosphate ceramics   72–73
chlorine anions   35
circulation efficiency   118–120
clay minerals   40, 42
collapse pressure, casing
  API plastic collapse pressure P_P 94
  API transition collapse pressure P_{tr} 92, 93
  API yield collapse pressure   94
  elastic collapse   92
colloidal dispersions of silica   46
compression waves   173
conductor casing   7, 8, 86
cone of AzNII   197, 198
consistometers   182–183
continuous intermediate casing   7, 8
continuous two-stage cementing   131
conventional jet mixer   81, 82
corrosion-resistant cement   65–66
C-90 steel grade   90, 91

d

diatomaceous earth (kieselgur)   44
dicalcium silicate (C_2S/belite)   19, 20
differential casing   99
dilatant   192
direct contact of reactants stage   20
dispersants
  definition   51
  plasticizers   51
  rheological properties of   50–51
  sedimentation   52–53
  superplasticizers   51
  water separation   52
displacement efficiency   118–120
displacement tank system   79

drill bit   1, 2
drilling
  intervals   5
  mud conditioning   120–122
  mud displacement   122–124
  mud displacement operations   117
  mud parameters   114
  operation   2, 4
  trip   5
  types of   3, 4
drill stem tester (DST)   167
drill string   1

e

elastic collapse   92
engineered particle-size distribution cements   67–69
equivalent cement sack mass   215
expansive cement   61–62
exploration wells   4
extenders   31, 41
  cement slurry density reduction of   39
  cement slurry volume, increase of   39
  clay minerals   40, 42
  extenders with low density   40
  gas-based   40, 48
  pozzolans   43–46
    diatomaceous earth (kieselgur)   44
    fly ash   44, 45
    lightweight cementing slurries   45
    lightweight particles   46–48
    silica (silicon dioxide, quartz)   45–46
  sodium silicate   43

f

FANN Instruments
  atmospheric consistometer   183
  cement slurry mixers   180
  HPHT consistometers   182
  MACS II multipurpose cement slurry analysis system   197
  pressurized mud balance   181
  ultrasonic cement analyze   200, 201
female plug, liner cementing   13
flapper-type check valve   99, 100
flexible cement   70–71

fluid loss   186, 187
fluid loss agents
  particulate materials   54
  water soluble polymers   54–55
fly ash   44, 45, 214–216
foamed cement   69–70
free water   52, 187–188
freeze-protected cement   62–63

g

gamma ray (GR) log   169–170, 176
gas-based extenders   40, 48
gas migration   202
gilsonite (asphaltum)   46–47
glass microspheres   47
GOST (Russian:ГОСТ) classification of Portland cement   29–30
granulated blast furnace slag   66
GR log *see* gamma ray (GR) log
guide shoe   97, 98
gypsum-Portland cement mix   63

h

hardware, casing
  casing shoe   96–99
  cementing basket   105, 106
  cementing head   104, 105
  cementing plugs   103–104
  centralizers   100–102
  check valve   99–100
  scratcher   103
  screening devices   105, 106
  turbulator   102, 103
hausmannite   49
heavy weight drill pipe (HWDP)   1
hematite   49–50
Herschel–Bulkley Model   192, 194–196
H-40 grade steel   89, 90
high-density cements slurries   69
high-pressure, high-temperature (HPHT) consistometers   182
high-speed formation   173
hinged centralizers   101
HPHT consistometers *see* high-pressure, high-temperature (HPHT) consistometers

HWDP *see* heavy weight drill pipe (HWDP)
hydraulic mixer   81–82
hydraulic testing
  inflow test   167
  pressure test   164–167
hydroxycarboxylic acid   38

i

ilmenite (iron titanium oxide)   49
injection wells   4
inorganic compounds   39
inorganic salts   32
intermediate casing   7, 8, 86
iron titanium oxide   49

j

J-55 grade   89, 90

k

kelly   1–3
K-55 grade   89, 90

l

laboratory evaluation of spacers and
     washers   204–205
Lamb wave   173
latex-cement systems   64–65
leak-off test   164, 165
L-80 grade steel   90, 91
lightweight cementing slurries   45
lightweight particles, pozzolans
  expanded perlite   46
  gilsonite (asphaltum)   46–47
  microspheres   47–48
  powdered carbon   47
lignosulfonates   37–38, 51
liner casing   7, 9, 87–88
liner cementing   13–14
  cementing head   135, 136
  hangers   134
  two liner plugs   135, 137
  types of   133
lipstick shoe   98
liquid additive dosing systems   80
liquid additive metering system   80
liquid sodium silicate   43

low-density cements
  foamed cement   69–70
  groups of cementing systems   69
low-density cement slurries   68

m

MACS II multipurpose cement slurry analysis
     system (FANN Instruments)   197
male plug, liner cementing   13
micro-annulus   35
microfine cements   71
microsilica (amorphous condensed
     microsilica)   45–46
microspheres
  ceramic   47–48
  glass   47
mixing fluid   180
mud conditioning   117
mud displacement
  circulation and displacement
       efficiency   118–120
  drilling mud conditioning   120–122
  drilling mud displacement   122–124
  drilling operations   117
  factor   117
  preparing the well for running casing   118
multiple pad sonic tool   177
multistage cementing
  continuous two-stage cementing   131
  standard two-stage cementing   128–130
  three-stage cementing   132

n

National Association of Corrosion Engineers
     (NACE)   89
natural pozzolans   43, 44
Newtonian fluids   192
N-80 grade steel   90, 91
nonaqueous cement systems   73
non-destructive test (ultrasonic
     measurement)   200, 201
non-Newtonian fluids   192, 193

o

oil well cementing technology development,
     history of   16–17

Index | 253

on-the-fly mixing   79
open hole remedial cementing equipment   108
organophosphonates   39
oxygen-activated neutron gamma method   171

p

perlite   46
plasticizers   51
plug cementing
   calculations   225–229
   equipment
      bridge plug   147–148
      diverter   148
      mechanical separators   148
      tailpipe or stinger   148
   evaluation   149
   slurry design   148–149
   techniques
      balance method   145
      cement plug installation using a dump bailer   145–146
      cement plug installation using the two plugs method   146
      cement plug installation with the use of coiled tubing   146–147
      definition of   144
plug flow   189
pneumatic loading (dry material)   75, 77
pneumatic mixing tank   75, 77
POD *see* point of departure (POD)
point of departure (POD)   183
polymelamine sulfonate   51–52
polynaphthalene sulfonate   51
poppet-type check valve   99, 100
pore pressure gradient (PPG) graph   5, 6
Portland cement   6
   API classification
      class A cement   25, 26
      class B cement   25, 26
      class C cement   25, 27
      class D cement   25, 27
      class E cement   25, 27
      class F cement   25, 28
      class G cement   25, 28
      class H cement   25, 28–29

chemical composition
   dicalcium silicate (C_2S/belite)   19, 20
   hydration process of   21–22
   tetracalcium alumoferrite (C_4AF/braunmillerite)   20
   tricalcium aluminate (C_3A)   20
   tricalcium silicate (C_3S/alite)   19, 20
definition   24
GOST (Russian: ГОСТ) classification of   29–30
manufacturing   22–24
powdered carbon   47
power-law model   192, 193
pozzolans
   diatomaceous earth (kieselgur)   44
   fly ash   44, 45
   lightweight cementing slurries   45
   lightweight particles
      expanded perlite   46
      gilsonite (asphaltum)   46–47
      microspheres   47–48
      powdered carbon   47
   silica (silicon dioxide, quartz)   45–46
PPG graph *see* pore pressure gradient (PPG) graph
precipitation theory   36
premixing liquid additives with water   79–80
pressure-integrity test (PIT)   164
primary cementing
   basket cementing   12–13
   calculation
      buoyancy   223–225
      buoyancy force   219
      hydrostatic pressures   222–223
      types of   218–219
      volume of cement slurry   221
      volume of displacing fluid   221
      well data   219–220
   cementing plugs   14, 15
   critical factors
      displacement of cement slurry   138–139
      volume of cement slurry   138
      well pressure   139–141
      well temperature   139, 140
   liner cementing   13–14, 133–137
      cementing head   135, 136

primary cementing (cont'd)
   hangers 134
   two liner plugs 135, 137
   types of 133
  mud displacement
   circulation and displacement efficiency 118–120
   drilling mud conditioning 120–122
   drilling mud displacement 122–124
   drilling operations 117
   factor 117
   preparing the well for running casing 118
  multistage cementing
   continuous two-stage cementing 131
   standard two-stage cementing 128–130
   three-stage cementing 132
  planning
   depth and design of the well 109–112
   drilling mud parameters 114
   reservoir conditions 113–114
  requirements 9
  reverse cementing 14, 15
  single-stage cementing with two plugs 10–11
  slurry selection
   cement slurry additives 116
   cement slurry design 116
   compressive strength and mechanical properties 115
   density 114–115
   formation temperature 115–116
  two-stage (two-cycle) cementing 11–12
  well cementing methods
   cementing through drill pipes 125–126
   cementing through small diameter (macaroni) tubing 126–127
   single-stage cementing 127–128
production casing 9, 86
production liner 87
production wells 4
pseudoplastic 192
pulsed neutron logging 170–171

q
"quick" method of Portland cement 23

r
radial acoustic cement meter 177
radioactive logging 169–171
reamer shoe 97, 98
recirculation hydraulic mixer 82–84
remedial cementing
  definition of 143
  equipment
   cased-hole 106–108
   open hole 108
  plug cementing
   calculations 225–229
   definition of 144
   equipment 147–148
   slurry design 148–149
   techniques 144–147
  squeeze cementing
   analysis and evaluation of 160–161
   applications 149–150
   calculations 229–236
   definition 144
   design and execution of 157–160
   filtration crust and cement nodes 150, 151
   slurry design 155–157
   technologies 152–155
   wellbore protrusions 151
  at the well-abandonment stage 143
  during the well-drilling stage 143
  during well operation 143
requirements 9
re-recorded sections 171
reservoir conditions
  pressure 113
  temperature 113–114
retarders 31
  adsorption theory 36
  cellulose derivatives 38
  hydroxycarboxylic acid 38
  inorganic compounds 39
  lignosulfonates 37–38
  organophosphonates 39
  precipitation theory 36

saccharide compounds   38
  theory of complexation   36
  theory of nucleation   36
reverse cementing   14, 15
rheological properties of cement slurry
  basic rheological concepts   190–191
  Bingham model   193–194
  flow types   188–189
  Herschel–Bulkley Model   194–196
  laminar flow   189
  Newtonian Fluids   192
  non-Newtonian fluids   192, 193
  power-law model   193
  turbulent flow   190
rigid centralizer   101, 102
Robertson and Stith's model   195
roller centralizer   101, 102
"Roman concrete"   22
rotary viscometer   195

S

saccharide compounds   38
salt-cement systems   63–64
scab liner   87, 88
scab tie-back liner   87, 88
scratcher   103
screening devices   105, 106
screw-type unloader   75, 77
"secondary" calcium hydrosilicate
    (C–S–H)   44
sedimentation, dispersants   52–53
self-orienting shoe   98
shear wave   173
shoe track   11
silica (silicon dioxide, quartz)   45–46
single-stage cementing with two
    plugs   10–11
slip-on centralizers   101
sodium chloride   32, 34
sodium salts   212–214
sodium silicate   43
sodium tetraborate decahydrate   39
solid sodium silicate   43
sound waves   172
special wells   5
specific gravity of cement slurry   208

squeeze cementing
  analysis and evaluation of   160–161
  applications   149–150
  calculations   229–236
  definition   144
  design and execution of
    cement slurry volume
      determination   157–158
    procedures for   159–160
    spacer, washer, and displacing
      fluids   158–159
    well injectivity determination   159
  filtration crust and cement nodes   150, 151
  slurry design
    fluid loss   156
    functions   155–156
    rheology   157
    thickening time   157
  technologies
    continuous injection of cement
      slurry   153, 154
    discontinuous injection of cement
      slurry   154
    squeezing pressure   152–153
    with the use of a cementing packer   155
    without using a cementing
      packer   154–155
  wellbore protrusions   151
standard two-stage cementing   128–130
steel grades, casing
  C-90 steel grade   90, 91
  H-40 grade steel   89, 90
  J-55 grade   89, 90
  K-55 grade   89, 90
  L-80 grade steel   90, 91
  N-80 grade steel   90, 91
  T-95 steel grade   90, 91
stop collar   99
storable cement slurries   73
storage silo   75, 76
strength characteristics of casing
  burst pressure   92
  collapse pressure   92–94
  yield strength   91
structural exploration wells   5
superplasticizers   51

surface casing 7, 8, 86
surface equipment
  cement trucks 78
  displacement tank system 79
  hydraulic mixer 81–82
  liquid additive dosing systems 80
  liquid additive metering system 80
  on-the-fly mixing 79
  pneumatic loading (dry material) 75, 77
  pneumatic mixing tank 75, 77
  premixing liquid additives with water 79–80
  recirculation hydraulic mixer 82–84
  screw-type unloader 75, 77
  storage silo 75, 76
  storage, transportation, and preparation of 75, 76
  surge tank 81
surge tank 81

t

temperature log 167–169
tetracalcium alumoferrite (C_4AF/braunmillerite) 20
theory of complexation 36
theory of nucleation 36
thermophysical properties of cement
  coefficient of linear thermal expansion 203, 204
  thermal conductivity 203
thixotropic cement 60–61
three-stage cementing 132
tie-back liner 9, 87
top drive 1
tricalcium aluminate (C_3A) 20
tricalcium silicate (C_3S/alite) 19, 20
triethanolamine ($N(C_2H_4OH)_3$) 36
T-95 steel grade 90, 91

turbulator 102, 103
two-stage (two-cycle) cementing 11–12

u

ultrafine Portland cements 63
ultrasonic tool 177–178
Ultrasound Cement Analyzer (UCA) 200
universal cementing unit 83, 84
U-tube effect 123–124

v

variable density log (VDL) 176

w

water soluble polymers 54–55
weighting additives for cement mortars
  barite 50
  hausmannite 49
  hematite 49–50
  ilmenite (iron titanium oxide) 49
  requirements 48
weighting agents 218
well cementing methods
  cementing through drill pipes 125–126
  cementing through small diameter (macaroni) tubing 126–127
  single-stage cementing 127–128
well construction
  definition of 7
  elements 1, 2
well drilling 1
wellhead 1, 2
wildcat well 5

y

yield strength 91